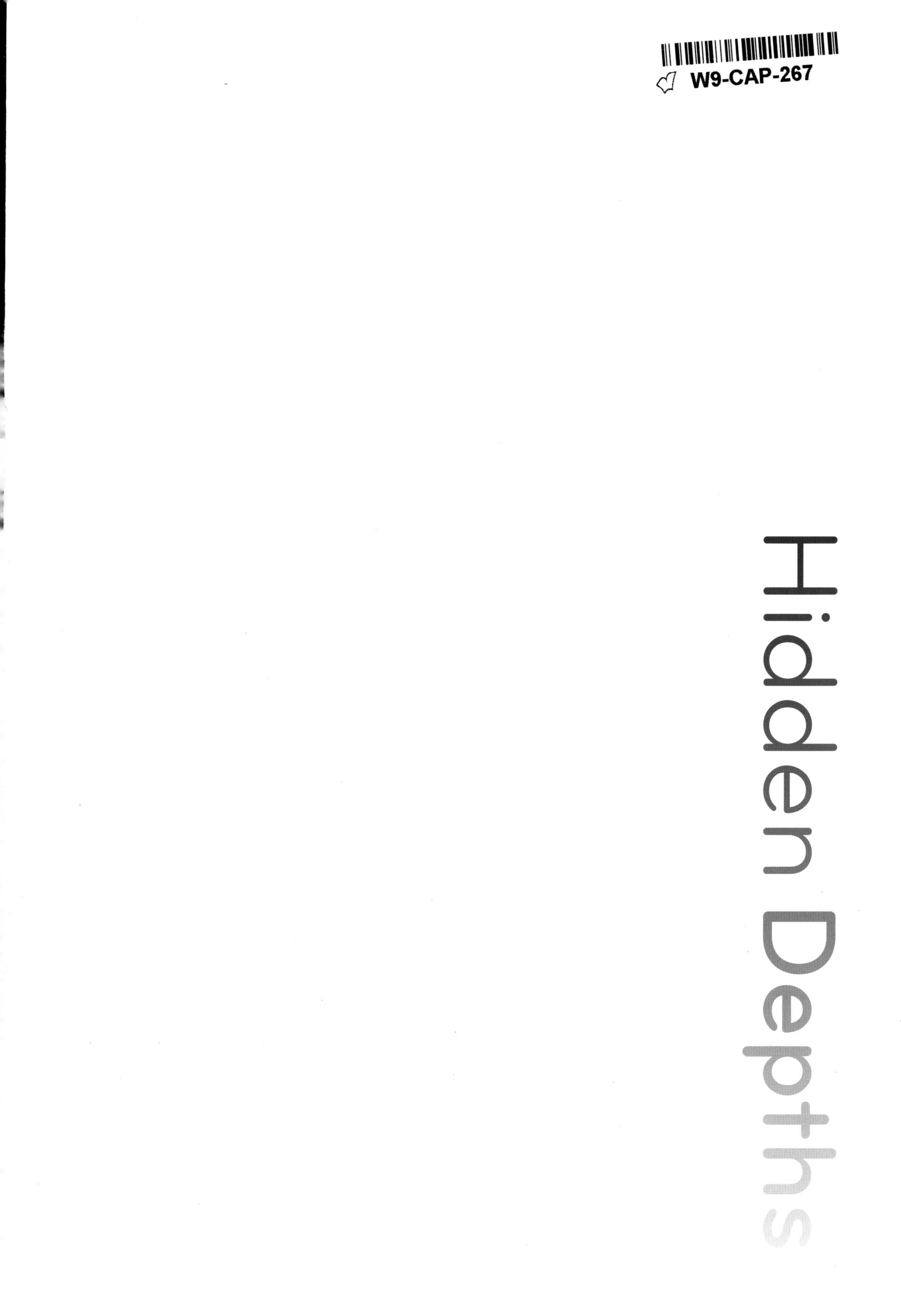
Hidden Depths

Hidden **Depths**
ATLAS OF THE OCEANS

10 East 53rd Street
New York
NY 10022
www.harpercollins.com

Published in collaboration with

The Smithsonian Institution,
Washington, D.C.
United States of America
www.si.edu

and

the National Oceanic and Atmospheric Administration (NOAA)
1315 East West Highway
Silver Spring, MD 20910
United States of America
www.noaa.gov

ISBN 978-0-06-134514-2

First Published 2007

Printed in Singapore

10 09 08 07
7 6 5 4 3 2 1

Hidden Depths

ATLAS OF THE OCEANS

Collins
An Imprint of HarperCollins*Publishers*

Published in collaboration with the
National Oceanic and Atmospheric Administration and the Smithsonian Institution

Contents

Foreword

Vice Admiral Conrad C. Lautenbacher, Jr., U.S. Navy (Retired) Undersecretary of Commerce for Oceans and Atmosphere and NOAA Administrator

Welcome to *Hidden Depths*, an atlas devoted to ocean education and literacy. This atlas is a product of a partnership developed between the National Oceanic and Atmospheric Administration (NOAA), and the Smithsonian Institution for the purpose of developing the Ocean Hall in the Smithsonian National Museum of Natural History. The goal of the Ocean Hall is to let visitors explore the oceans through exhibits, images, and videos. This atlas is a companion volume to the Ocean Hall which will bring the wonders of the ocean into homes worldwide.

More than 70 per cent of the surface of the Earth is covered by oceans. The oceans moderate the earth's climate, provide a source of food, and nurture tiny plant-like creatures which are the source of 50 per cent of the planet's oxygen. Since the dawn of time, people have viewed the seas as a source of spiritual renewal and inspiration. That being said, we know more about the surface of the Moon than the floor of the ocean.

Scientists and explorers have studied the oceans for centuries. We have increased our understanding of the ocean, including relationships between living creatures and their ecosystems, the ocean's role in weather and climate, biological resources for food and medicine, and the dynamic ocean seafloor where tectonic plates slide past one another, spread apart, or are subducted, one plate under another. But, by and large, the oceans remain a largely unexplored and unobserved area, the last great frontier on our blue planet where many discoveries remain to be made.

We depend on our oceans for our way of life. Unfortunately, that dependence has placed pressure on this vital resource. Some fisheries stocks have been depleted; pollution from river run-off and intentional dumping has affected large regions of the coastal ocean; increased maritime activities have stressed coastal areas; and increased carbon dioxide has led to an increase in the acidity of our oceans.

Many positive developments are counteracting these forces. There are thousands of marine protected areas around the world, and this was highlighted by the recent establishment of the Papahanaumokuakea Marine National Monument in Hawaii, the largest marine conservation area on earth. The U.S. government has committed to end overfishing by 2010 and is working internationally to end destructive fishing practices on the high seas. Across the globe, there is a renewed understanding that we all must be good stewards of this resource which brings us so many benefits.

Undoubtedly there will be many challenges in the future. The global population is projected to increase by 50 per cent. Demographic trends project that much of this growth will be in urban centers and coastal areas, placing even more stress on ocean resources. Furthermore, increased populations and development in coastal regions place more people in the path of natural disasters like tsunamis, hurricanes, and cyclones. To meet these challenges, we must improve our understanding of the complex anatomy of Earth's many systems.

To this end, the United States is working aggressively with dozens of other countries and non-governmental organizations to build a Global Earth Observation System of Systems (GEOSS)

which will monitor the complete Earth environment. This system will facilitate a free flow of data and information from all corners of the world to scientists and policy makers to enhance our ability to diagnose and forecast changes in the Earth's environment. This increased knowledge could allow us to improve forecasts of hurricane landfall several days in advance, of how a disease outbreak might spread or where a farmer might plant certain types of crops to maximize yield.

The oceans are integral to many of the planet's systems. Therefore, to successfully develop GEOSS, we must start with the oceans. Efforts are underway to develop integrated ocean observing systems which combine data from buoys, satellites, fixed tide gauges and water sampling installations, and research ships to provide a complete picture in real time of the state of our oceans. Integrating these observations will help us provide more accurate advanced forecasts of climate and weather, improve the safety and efficiency of maritime operations, mitigate the effects of natural hazards, and protect and restore healthy coastal ecosystems.

Integrated observations mark a break from the past. Historically, scientists studied the oceans in a piecemeal approach by looking at a single species or area. Today scientists understand that the entire ocean is linked and to gain understanding of one component of it they must have a complete picture of the entire system, including the interplay between the physical, biological, chemical, and geological processes in the ocean.

It is an exciting time to be involved in ocean policy, ocean exploration and ocean research. The oceans regularly make front-page news and they have become a top priority at the highest levels of government. Everyone shares the common understanding that the oceans are vital to our way of life and therefore we must be good caretakers of the resource.

Ensuring that we are good stewards of the ocean requires an informed and ocean-literate citizenry. This atlas, *Hidden Depths*, and the Smithsonian's new Ocean Hall are significant steps towards fulfilling NOAA's commitment to furthering lifelong ocean education and developing future generations of scientists and policymakers committed to ocean stewardship.

The National Oceanic and Atmospheric Administration (NOAA) is a scientific agency focused on the conditions of the ocean and atmosphere. NOAA and its ancestor agencies have charted seas and skies, warned of dangerous weather, guided the use and protection of ocean and coastal resources, and conducted crucial research in improving understanding and stewardship of the environment.
This book has been authored by many NOAA scientists whose specialisms cover a broad spectrum of knowledge and scientific endeavor within this field.

Atlas of the oceans

Atlas of the oceans
Introduction

The water of the world's oceans covers nearly three-quarters of the earth's surface, and exerts an extraordinary influence on the physical processes of the earth and its atmosphere. The circulation of water throughout the oceans is critical to world

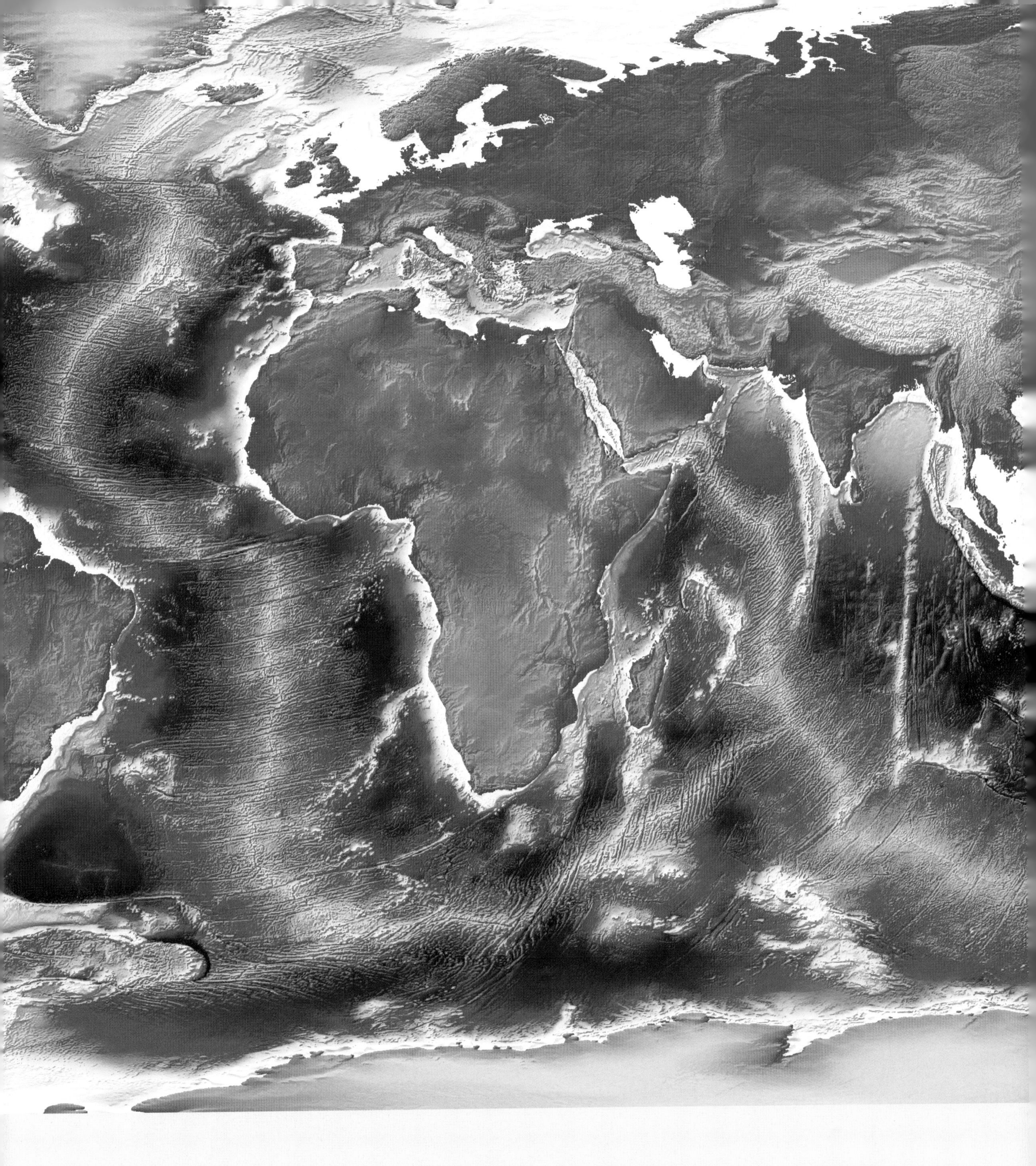

climate and climate change, so much so that any study of the relationships between such aspects of the earth and its climate relies upon a clear understanding of the role of the oceans and of the complex processes within them. The precise nature and effect of the processes varies geographically, as demonstrated by the El Niño phenomenon, but individual instances cannot be considered in isolation from the overall processes involved.

ABOVE:
Global sea floor topography from satellite altimetry

The physical world

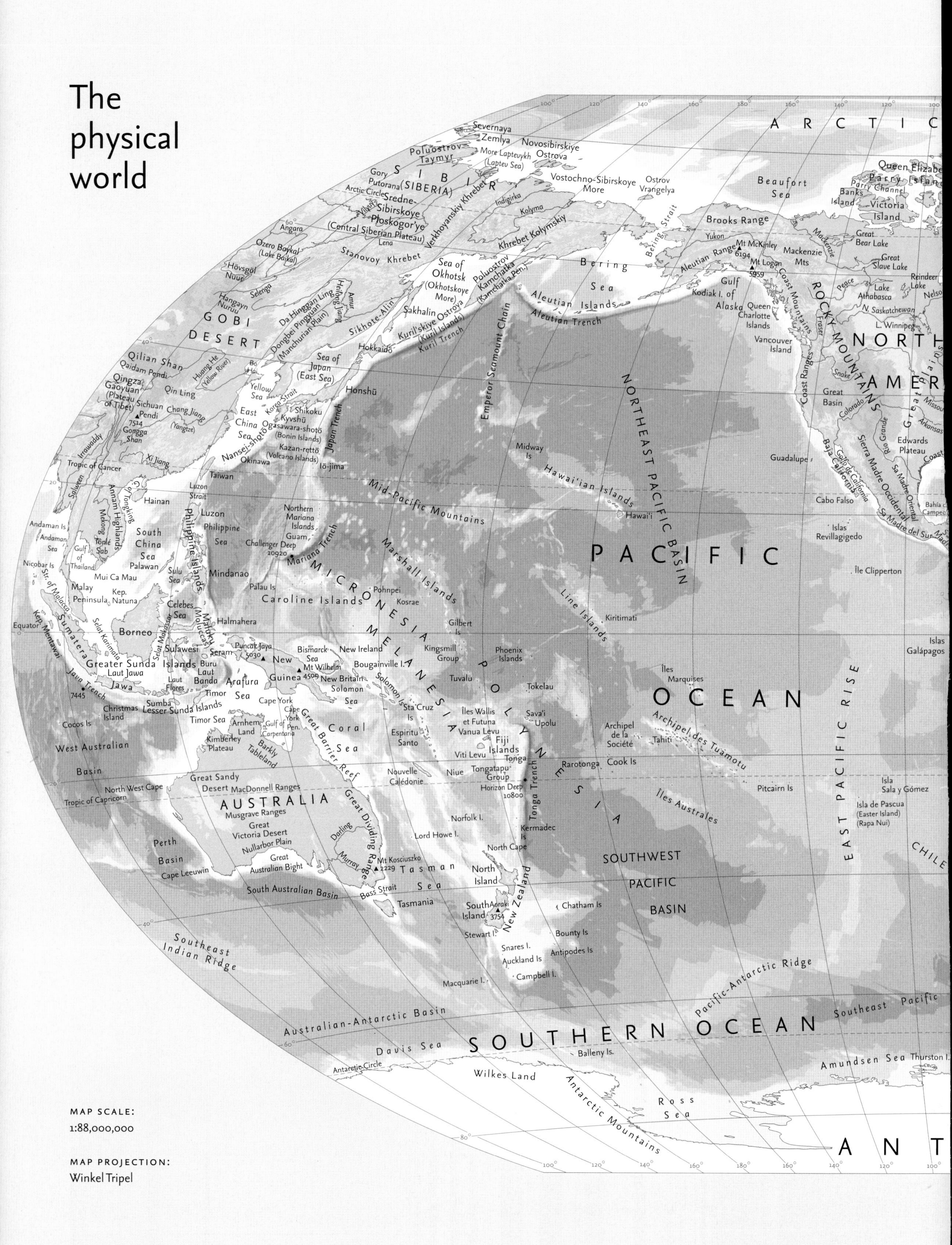

MAP SCALE:
1:88,000,000

MAP PROJECTION:
Winkel Tripel

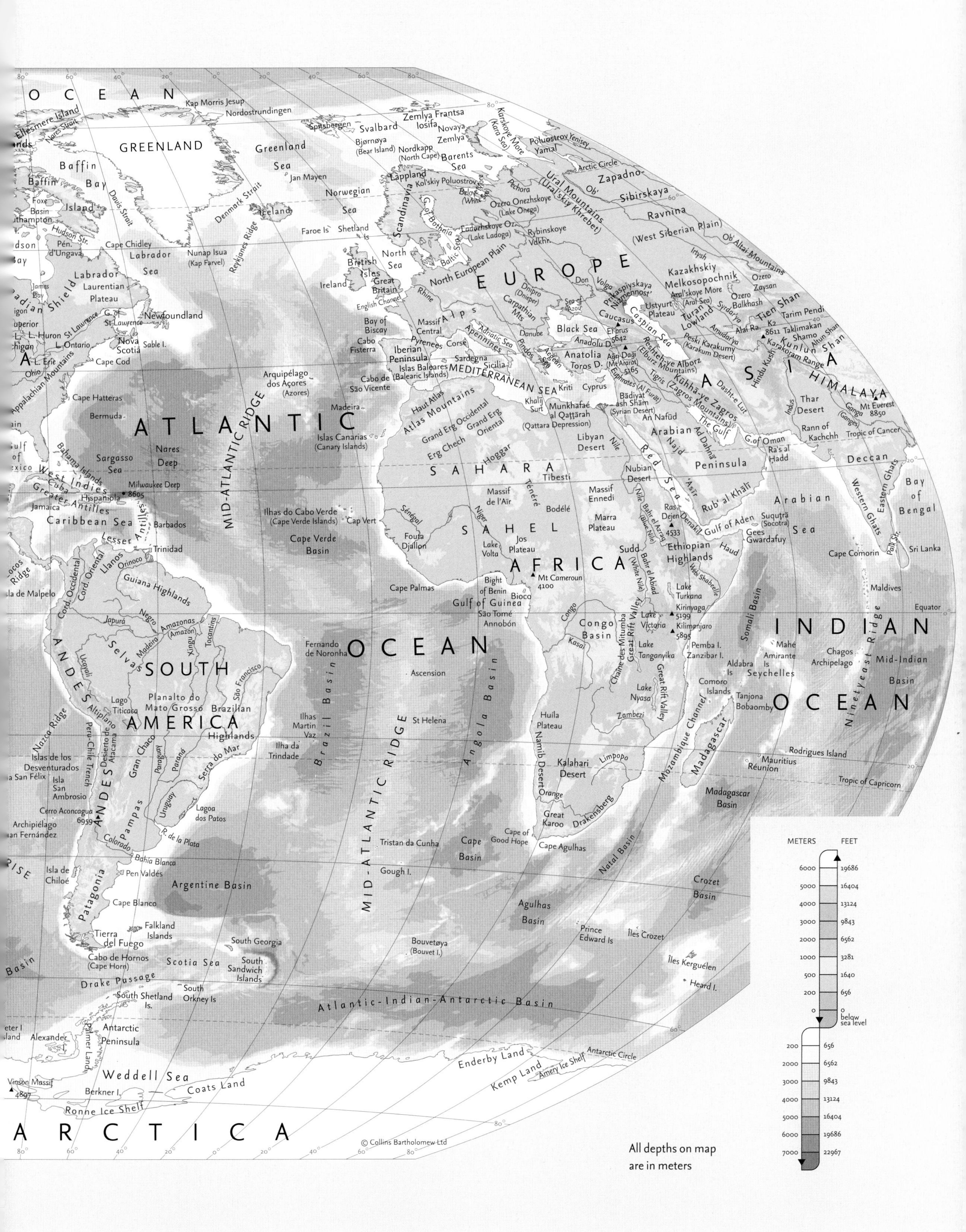

All depths on map are in meters

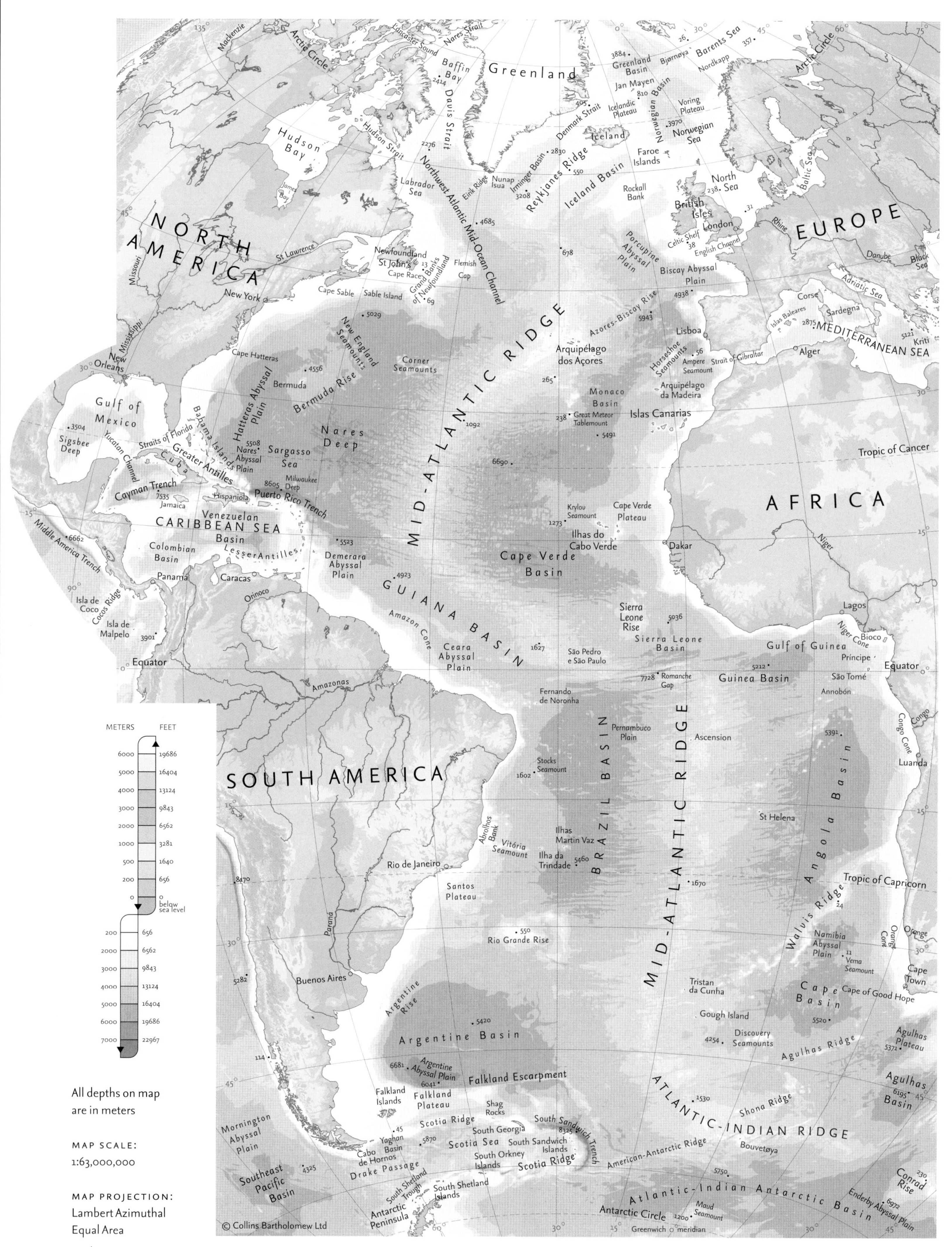

All depths on map are in meters

MAP SCALE:
1:63,000,000

MAP PROJECTION:
Lambert Azimuthal Equal Area

World Ocean

The five Oceans of the world, Arctic, Atlantic, Indian, Pacific, and Southern cover over 70 per cent of the surface of the earth. At any one time approximately 96 per cent of the earth's water - around 310 million miles^3 (1,300 million km^3), is held within their depths. As easily identifiable distinct bodies in their own right, they are often thought of separately, however they are really one group of interconnected water bodies which come together to form one continuous salt water entity. This entity is sometimes referred to as the Global or World Ocean.

All of the water on the planet when taken together constitutes the earth's hydrosphere. In taking up such a large percentage of this total, the World Ocean must be considered the most important component of the earth's hydrosphere and as such is a massive driver of the planet's climate and weather patterns. These in turn affect all forms of life on the earth.

Even though their influence on the planet is huge, the science of oceanography, or the study of the oceans, is relatively new in historical terms. The explorers of the middle ages can be considered the first oceanographers but it was not until the eighteenth and early nineteenth centuries that scientific texts on the subject began to be written and sounding techniques were used to accurately map the ocean floors.

Today, direct and indirect sampling of the oceans through satellite remote sensing and radiometry, for example has allowed the science to progress very rapidly. Developments in these fields mean that the ocean floors can now be mapped with precise detail.

The next few pages focus on the three major oceans on the earth, the Atlantic, the Indian, and the Pacific. For more information and maps of the Arctic and Southern Oceans, see the Poles section (pages 118-133).

PREVIOUS PAGE:
Physical map of the world. The oceans of the world cover over 70 per cent of the total surface area of the planet.

Atlantic Ocean

The second largest ocean on earth is the Atlantic Ocean stretching from the Americas in the west to Europe and Africa in the east. Sometimes divided into northern and southern sections the Atlantic connects with the Arctic Ocean in the north, the Indian in the southeast, the Pacific in the southwest, and the Southern Ocean at its southern most edge.

Taken in isolation the Atlantic occupies an area of around 31.8 million sq miles (82.4 million sq km), although when adjacent seas such as the Caribbean and the Mediterranean are added to this the final total is often considered to be nearer 41.1 million sq miles (106.4 million sq km), an area representing approximately 20 per cent of the earth's surface.

The deepest point in the Atlantic at 28,231 ft (8,605 m) below sea level, is found in the Puerto Rico Trench to the north of the island of the same name. The key feature of the Atlantic's floor is not found here however, but in the form of a huge submarine mountain range known as the Mid-Atlantic Ridge. This ridge and the associated rift valley found at its core, runs from Iceland in the north down to a latitude of 58°S, making it one of the earth's longest mountain ranges.

RIGHT:
Satellite altimetry image depicting the Atlantic Ocean floor in a range of colours to show depth, orange is the shallowest with blue and purple being the deepest. Clearly visible is the Mid-Atlantic Ridge shown in green and yellow.

LEFT:
Map of the Atlantic Ocean.

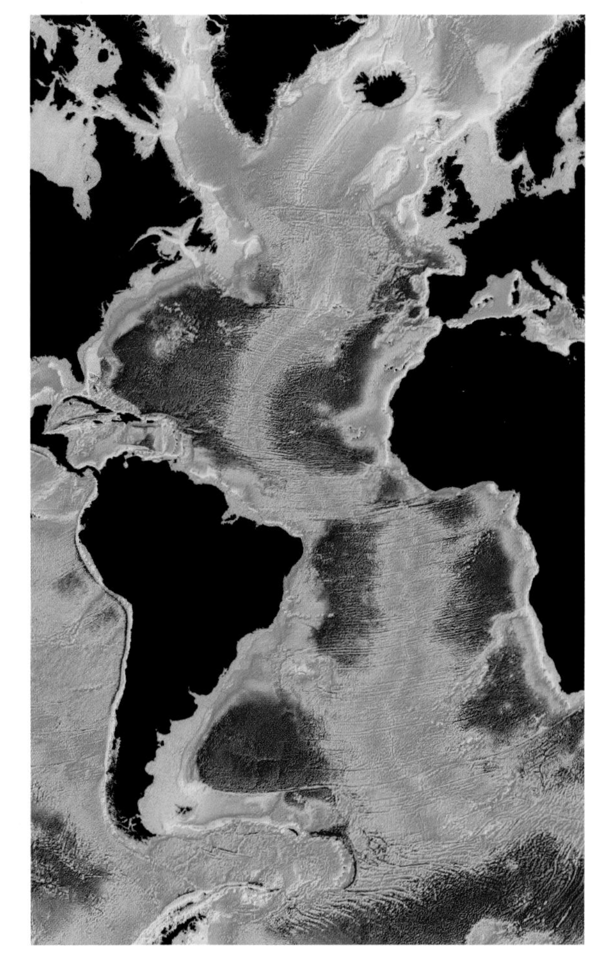

Indian Ocean

The Indian Ocean at 28.35 million sq miles (73.43 million sq km) is the earth's third largest ocean behind the Pacific and Atlantic. Lying in the gulf created by Africa to the west, Asia to the north and Oceania to the east it has the least northern hemisphere extent of the three major oceans and does not join with the Arctic Ocean at all. The southernmost boundary of the Indian Ocean is at the point where it meets the Southern Ocean at around 60°S. As with the other oceans of the world the Indian also encompasses several major seas within its parameters, most notable of which are the Arabian Sea and the Red Sea.

Unlike the Atlantic Ocean which is bisected almost perfectly by one major ridge, the Indian Ocean floor is criss-crossed by a series of ridges and plateaux The deepest point of the Indian Ocean at 24,425 ft (7,445 m) is found in the Java Trench, south of the island.

RIGHT:
Satellite altimetry image depicting the Indian Ocean floor. The blue areas of the deep sea are broken up by a patchwork of greens and oranges representing the shallower ocean floor ridges.

FAR RIGHT:
Map of the Indian Ocean.

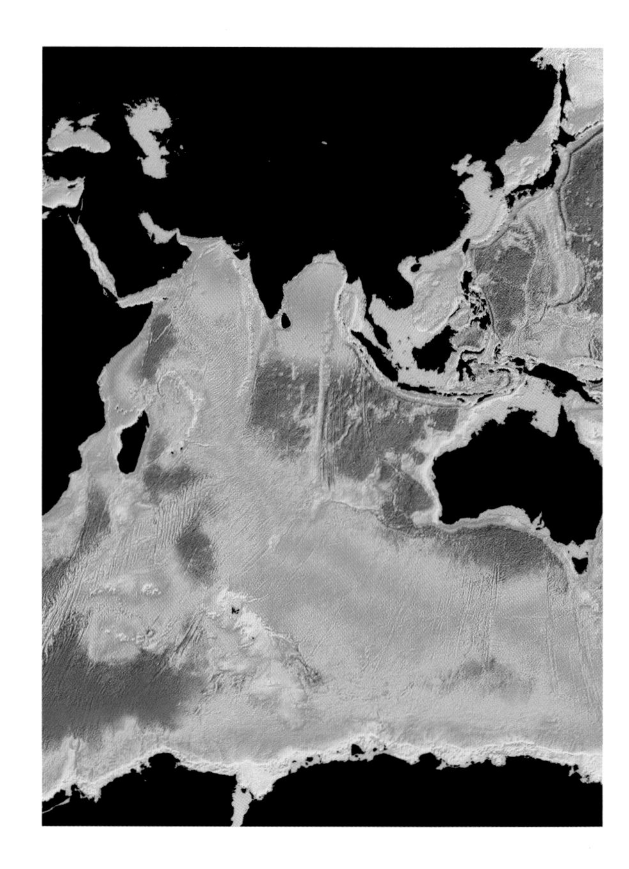

Pacific Ocean

The largest ocean on the planet is the Pacific which covers around 32 per cent of the earth's surface area at 64.19 million sq miles (166.24 million sq km). It is bounded by the Americas to the east and northeast, Asia to the northwest and west, as well as Australia to the southwest. The southern edge of the Pacific opens into the Southern Ocean. Like the Atlantic, the Pacific also meets the Arctic Ocean at its northern extreme in the form of the 58 miles (92 kilometers) wide Bering Strait.

The deepest point in the Pacific Ocean is the bottom of the Mariana Trench which is around 6.77 miles (10.92 km) deep. Not only the deepest point in the Pacific this is also the single deepest point on the earth's crust, located to the south and east of the Mariana Islands and Guam at a subduction zone between two tectonic plates, the Pacific (subducted) and the Philippine. The Mariana is one of a number of deep subduction zone trenches located towards the ocean's perimeter, the Japan and Kuril trenches being other notable examples, a fact which explains why the areas bounding the Pacific (sometimes called the Pacific "Ring of Fire,") experiences more geological activity than almost any other region.

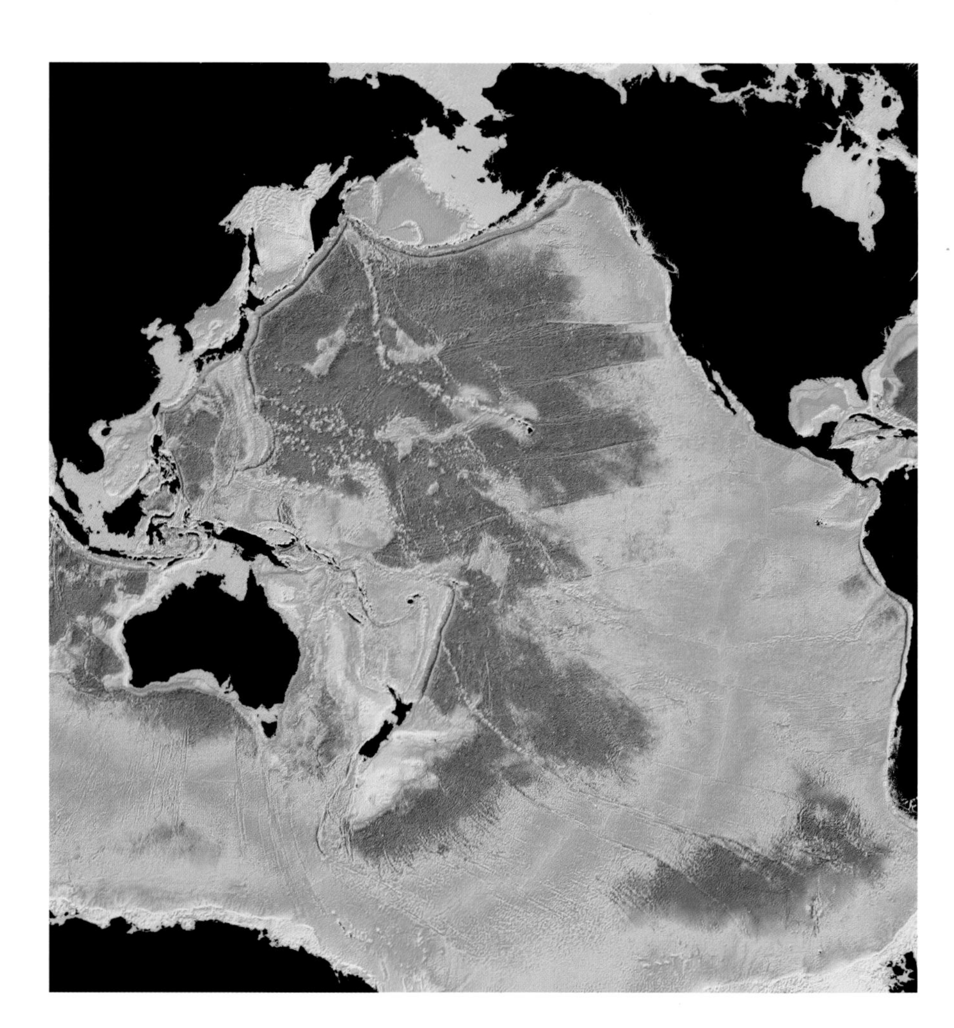

PAGES 18-19:
Map of the Pacific Ocean.

ABOVE:
Satellite altimetry image depicting the Pacific Ocean floor. Orange and yellow represent the higher floor with the deepest trenches , off the coast of Japan and north of New Zealand shown as purple.

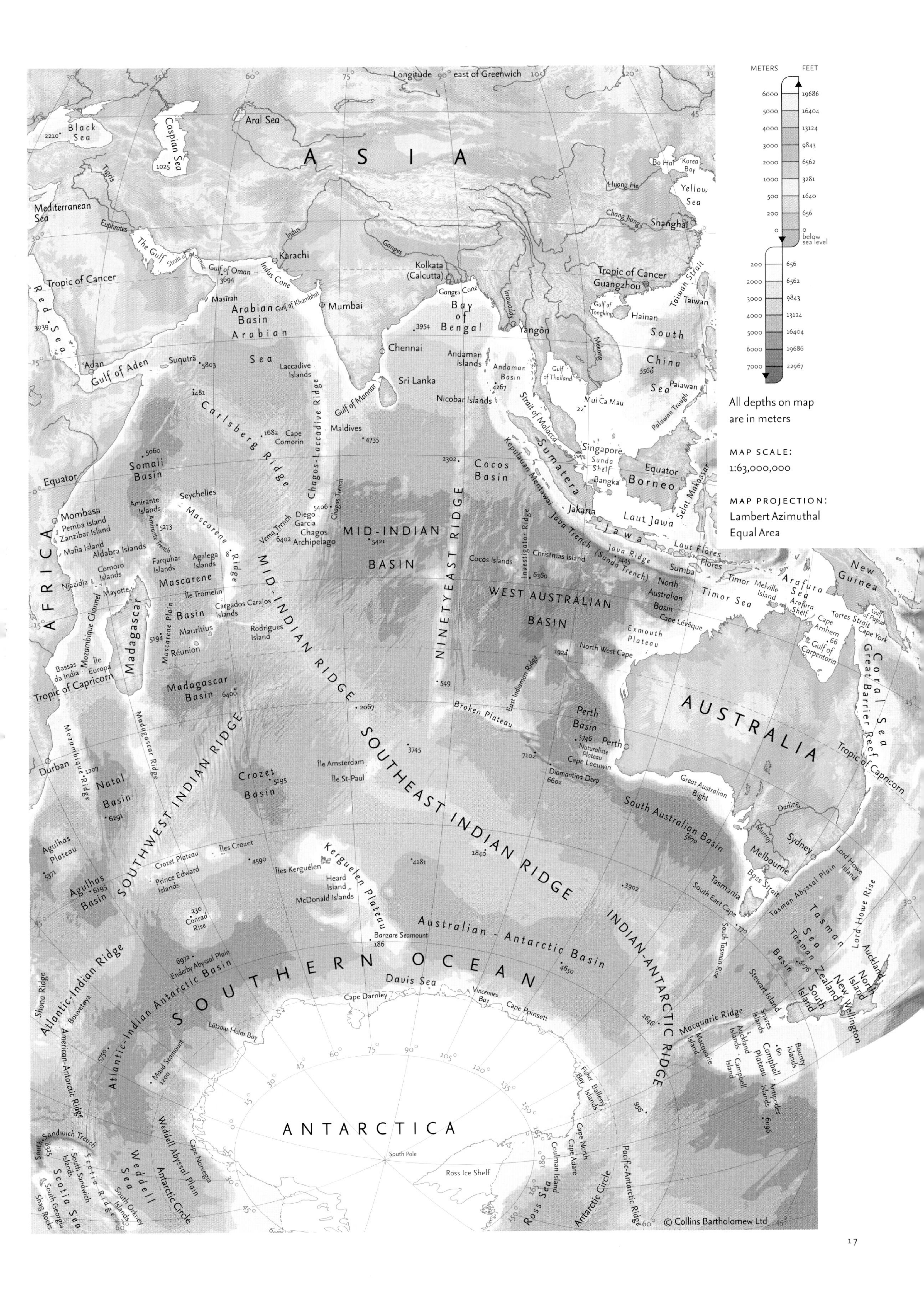
METERS
FEET
All depths on map are in meters
MAP SCALE:
1:63,000,000
MAP PROJECTION:
Lambert Azimuthal Equal Area
ASIA
AFRICA
AUSTRALIA
ANTARCTICA
SOUTHERN OCEAN
Arabian Sea
Bay of Bengal
MID-INDIAN BASIN
NINETYEAST RIDGE
SOUTHEAST INDIAN RIDGE
SOUTHWEST INDIAN RIDGE
MID-INDIAN RIDGE
WEST AUSTRALIAN BASIN
Australian - Antarctic Basin
INDIAN-ANTARCTIC RIDGE
Longitude 90° east of Greenwich
© Collins Bartholomew Ltd

Pacific Ocean

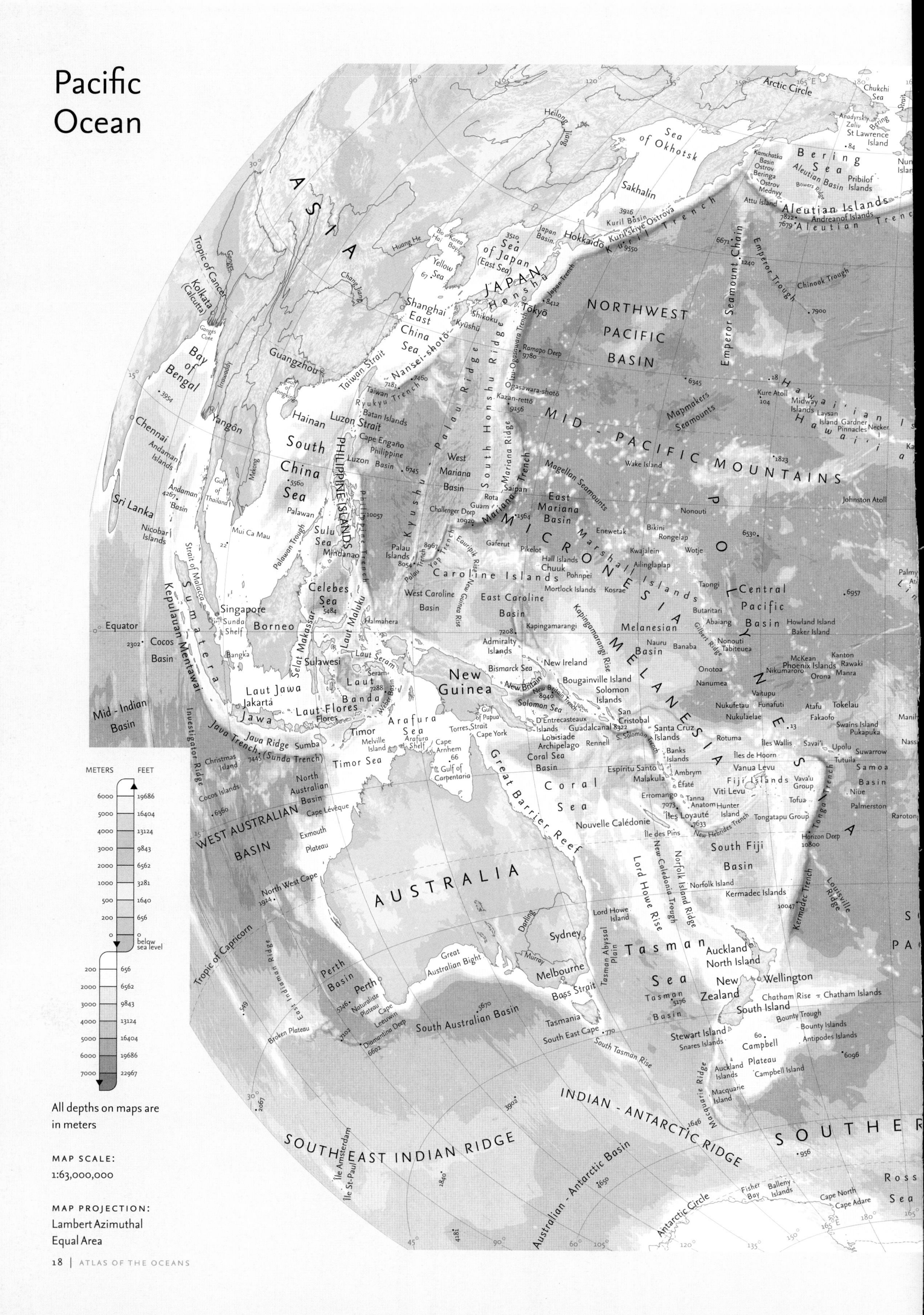

METERS | FEET

METERS	FEET
6000	19686
5000	16404
4000	13124
3000	9843
2000	6562
1000	3281
500	1640
200	656
0	0 below sea level
200	656
2000	6562
3000	9843
4000	13124
5000	16404
6000	19686
7000	22967

All depths on maps are in meters

MAP SCALE:
1:63,000,000

MAP PROJECTION:
Lambert Azimuthal Equal Area

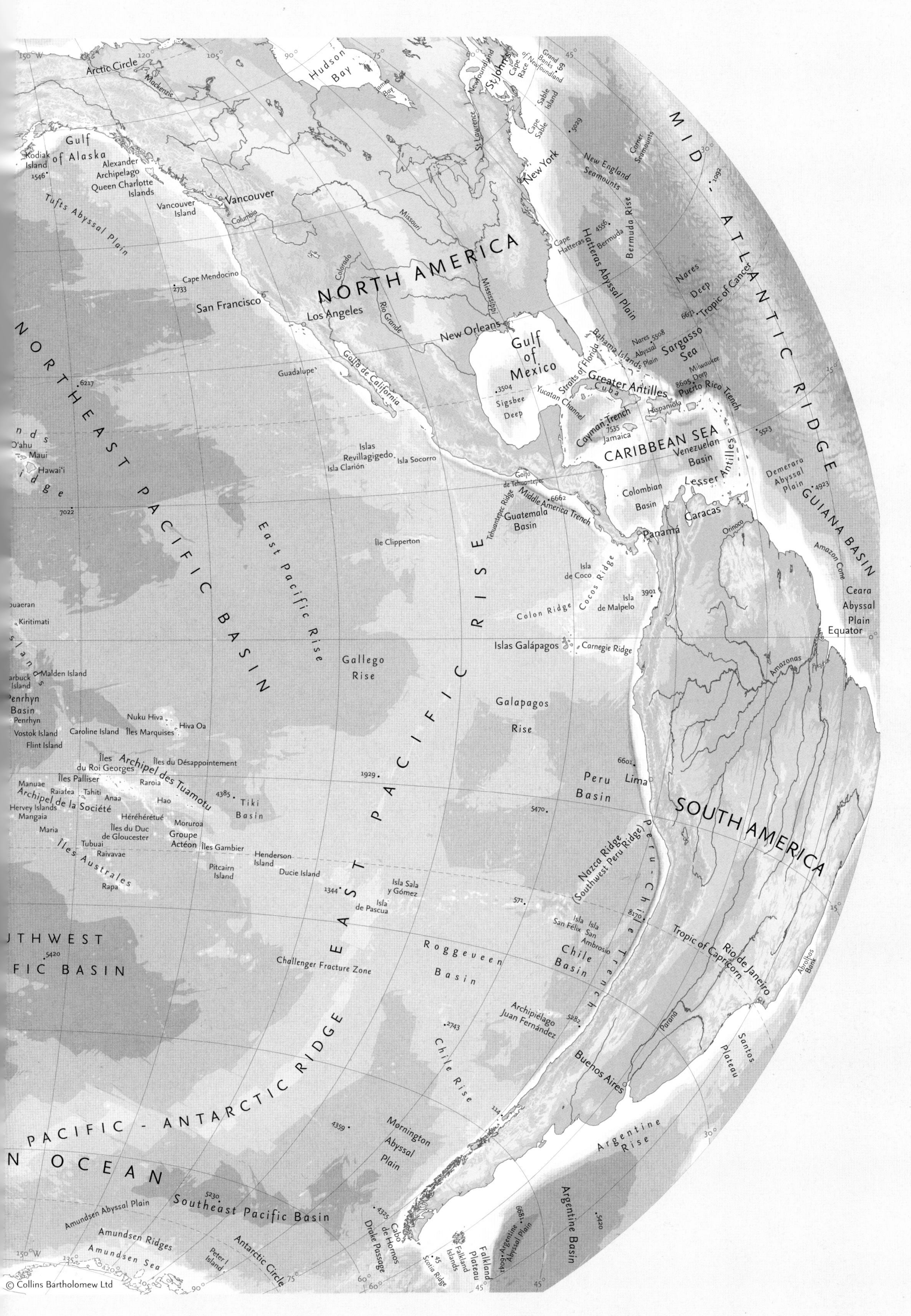

NORTH AMERICA
SOUTH AMERICA
NORTHEAST PACIFIC BASIN
EAST PACIFIC RISE
PACIFIC - ANTARCTIC RIDGE
MID - ATLANTIC RIDGE
CARIBBEAN SEA
Gulf of Mexico
Hudson Bay
Gulf of Alaska
Sargasso Sea
Greater Antilles
Lesser Antilles
GUIANA BASIN
Southeast Pacific Basin
Argentine Basin
Peru Basin
Chile Basin
Tiki Basin
Roggeveen Basin
Galapagos Rise
Gallego Rise
Chile Rise
Argentine Rise
Mornington Abyssal Plain
Vancouver
San Francisco
Los Angeles
New Orleans
New York
Caracas
Panamá
Lima
Buenos Aires
Rio de Janeiro
Equator
Tropic of Cancer
Tropic of Capricorn
Arctic Circle
Antarctic Circle

The ocean floor

The Ocean floor

Introduction

Earth's ocean covers more than 70 per cent of its surface and the seafloor "resides" at an average depth of 9,900 to 13,000 ft (3,000 to 4,000 m) below this. The continents' elevation is, on average, approximately 2,750 ft (840 m) above sea level. This "bimodal" division in the earth's topography is caused by the density difference between the relatively dense volcanic rocks of the ocean floor (basalts), and the lighter rocks of the continents (andesites and granites). The oldest ocean floor is only about 170 million years old, as opposed to 4 billion years for the oldest known continental rocks (a site in Canada). In a very broad sense, the long-term cooling of the earth from its molten birth more than 4.5 billion years ago has produced a differentiation of lighter material (continents) from denser material. However, it wasn't until the development of the theory of plate tectonics, the grand unifying theory of earth science, in the 1960s that we came to better understand the process behind the basic observations of the ocean's youth and its volcanic foundation.

The earth's surface is constantly in motion. Its outer crustal shell (lithosphere) is divided into a series of more or less rigid plates which move relative to one another, driven by slow-moving convection cells in the earth's upper mantle (asthenosphere). We don't feel this motion at the interior of these plates – it's mostly at the plate boundaries where the geologic energy (ultimately derived from the still-hot interior of the planet) is released as plates collide with, slide past or separate from one another. Thus, a map of earthquake and volcano locations outlines the plate boundaries. There are fourteen large plates which encompass most of the earth's surface (including almost the entire ocean floor) and thirty-eight smaller ones. The United States and Canada are part of the North American (NAM) plate, encompassing the entire North American continent from the San Andreas Fault on the west (where the North American plate slides past the Pacific plate) to the Mid-Atlantic Ridge on the east. The NAM plate is composed of both continental and oceanic sections because there is no plate boundary on the east coast of North America. The largest plate is the Pacific plate (PA). Its eastern boundary is the San Andreas Fault, its northern boundary is the Aleutian Trench, its southern boundary is the Pacific-Antarctic Ridge, and the western boundary is a series of ridges and trenches (for example the Mariana Trench).

The plates have a finite life cycle. Plates are born as new magma from the earth's interior rises up into cracks and solidifies into basaltic rock along the Mid-Ocean Ridge which stretches for more than 37,000 miles (60,000 km) through the ocean basins. The undersea basaltic lava flows form characteristic pillow shapes (see page 25) as they are extruded to make new ocean floor. This new ocean crust spreads away from the ridge in both directions at rates averaging up to about 3 in (7.5 cm) per year. As it cools, it contracts and slowly deepens as it moves away. The oldest ocean floor in the western Pacific is more than 20,000 ft (6,000 m) in depth before it begins the descent into the Mariana Trench. The ocean plates are recycled as they descend beneath other plates into the great ocean trenches and subsumed back into the interior of the earth over millions of years. These subduction zones are complex and include both the deepest parts of the ocean (trenches) and some of the shallowest submarine volcanoes.

In some places plate collisions push up new mountain systems such as the Himalaya. The relative youth of the oceans implies that the entire ocean floor has been renewed several times since plate tectonics began at least one billion years ago. Many of the mountain belts of the world (for example the Appalachian) were formed during previous cycles of plate

PREVIOUS PAGE:
Underwater boat searching for living fossils in the Bahamas, Caribbean Sea. The fossils are stromatolites, composed of sticky mats of cyanobacteria which cement sand and sediments. They appear in rocks 3.5 billion years old.

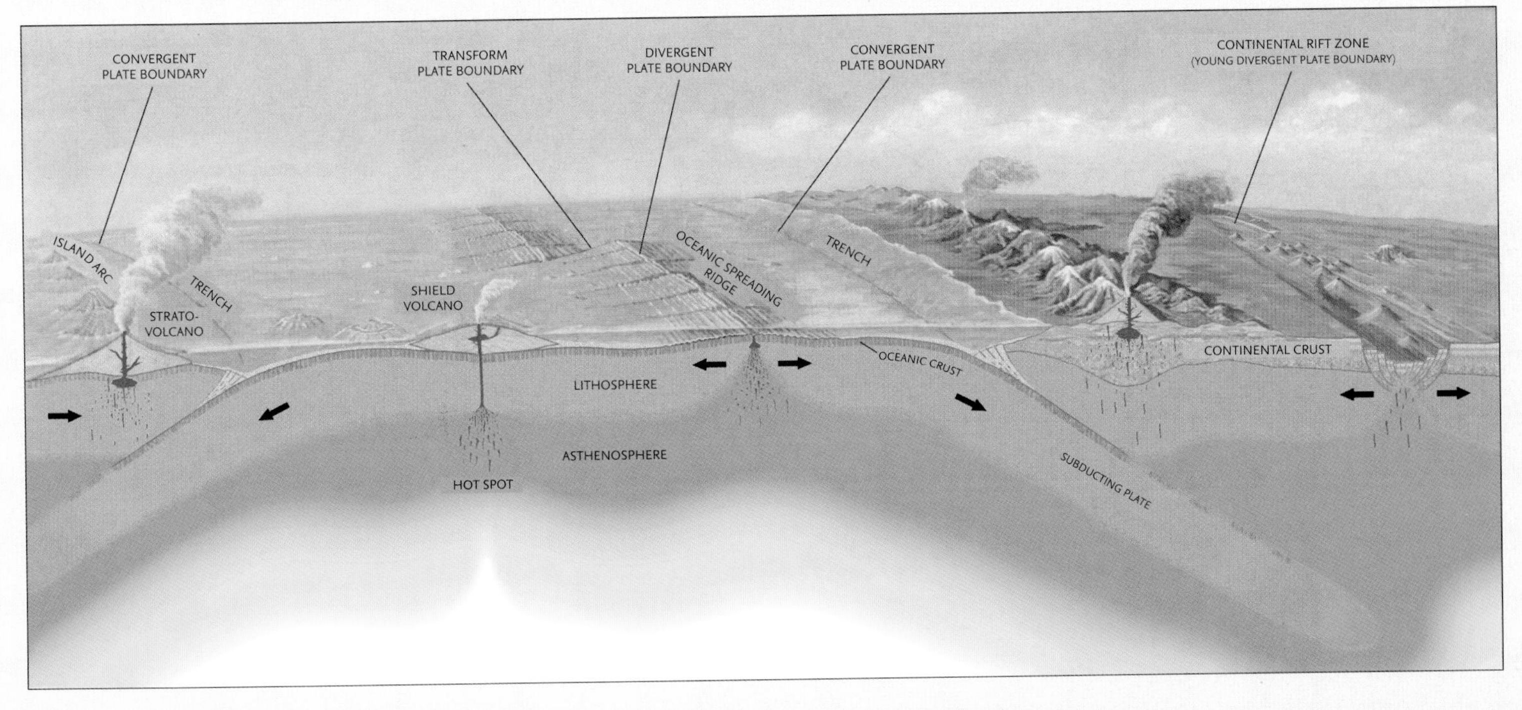

tectonics. In any case, it is a zero-sum process, because the earth is not expanding.

Although virtually the entire ocean floor is underlain by basaltic lavas, most of these have been covered by a relatively thin veneer of sediments accumulating since the ocean crust was born at the Mid-Ocean Ridge. However sediments can attain thickness of several kilometers at sites near the continents where deposition from river runoff is high. Sediments in the interior of the ocean basins mostly consist of slower-accumulating fine clays and shell material deposited from above. Outcropping of volcanic rock is mostly confined to the proximity of the Mid-Ocean Ridge and to high-standing features such as seamounts.

ABOVE:
Artist's cross section illustrating the main types of plate boundaries.

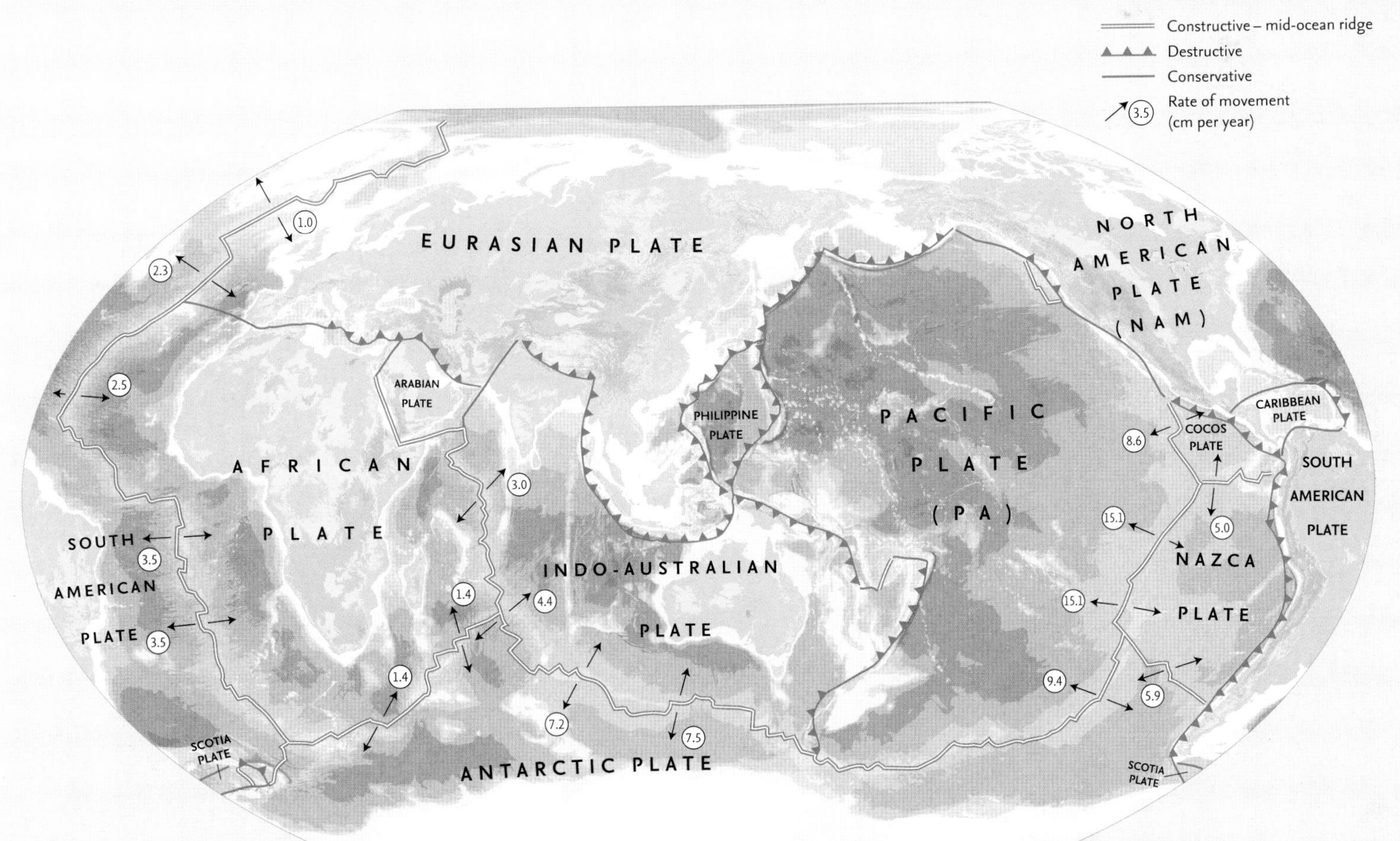

Submarine volcanic activity

The major growth boundary of the earth's tectonic plates is the earth-girdling Mid-Ocean Ridge (MOR). This system is almost entirely submarine with only small portions above sea level, such as Iceland. Episodic volcanic activity occurs on hundreds of "stretched" volcanoes (because the boundary is under tension) along the MOR. This volcanic activity is almost entirely at depths greater than 6,500 ft (2,000 m). The character of the MOR changes markedly according to the rate of spreading and the amount of magma delivered to the surface over a given period of time. At the faster spreading rates (for example in the eastern Pacific) an elongate ridge is present and volcanic activity is more frequent. At the slower spreading rates (and/or low supply of magma) there is usually a distinct "rift valley" (for example the Mid-Atlantic Ridge) and volcanic eruptions are less frequent.

Other types of submarine volcanic activity have formed many of the large undersea mountains called seamounts and the broad oceanic plateaux found in all the ocean basins. Some of these features have formed above hotspots—the surficial expression of deep mantle "plumes" rising to the earth's surface over many millions of years. Chains of seamounts often form as the plate moves over the hotspot (as in the Hawaiian Islands). The larger anomalously shallow ocean plateaux are probably the result of very high rates of ocean volcanic activity driven by "superplumes" at times in the past.

In contrast to the Mid-Ocean Ridge, the submarine volcanoes formed in association with the oceanic trenches (subduction zones) have summit depths which average less than 6,500 ft (2,000 m). These volcanoes occur above the earth's great subduction zones (trenches) where gases (for example carbon dioxide) released during the sinking and heating of the oceanic plate migrate upward to precipitate melting of the overlying mantle. Volcanic activity above these melt zones—commonly 90–120 miles (150–200 km) from the trench—builds composite cones made up of interlayered ash and lavas. The Cascade volcanoes of Oregon and Washington are good examples of this subduction-related "island arc" volcanism where an oceanic plate sinks below a continent (here the trench is buried with sediment). In the case where an old ocean plate sinks below another ocean plate, the submarine volcanoes grow into shallow water with some emerging as islands (for example the Mariana arc). These volcanic arcs can be traced as a band around the west coasts of the Americas, through the Aleutian Islands and then southwards through Japan, the Philippines and around to the west and south through the southwest Pacific. This is the circum-Pacific "Ring of Fire." The 1,700 mile (2,800 km) system of parallel trench and arc volcanoes lying between Japan and the island of Guam has many active island volcanoes and even more numerous submarine volcanoes. Another long chain of island arc submarine volcanoes and islands occurs north of New Zealand. Shorter arcs are found around Samoa, New Guinea and on the north and west sides of the Philippine Sea. The lavas produced along the volcanic arcs have, on average, more silica and more gases than hotspot or MOR lavas. This compositional difference (along with the shallower average depth of their summits) makes them more explosive than MOR eruptions.

Recent expeditions to the Mariana and Kermadec arcs have discovered some very exciting new sites, including one which has been erupting for a couple of years, another with liquid carbon dioxide emissions and sites where liquid sulfur exists on the seafloor under more than 40 atmospheres of pressure .

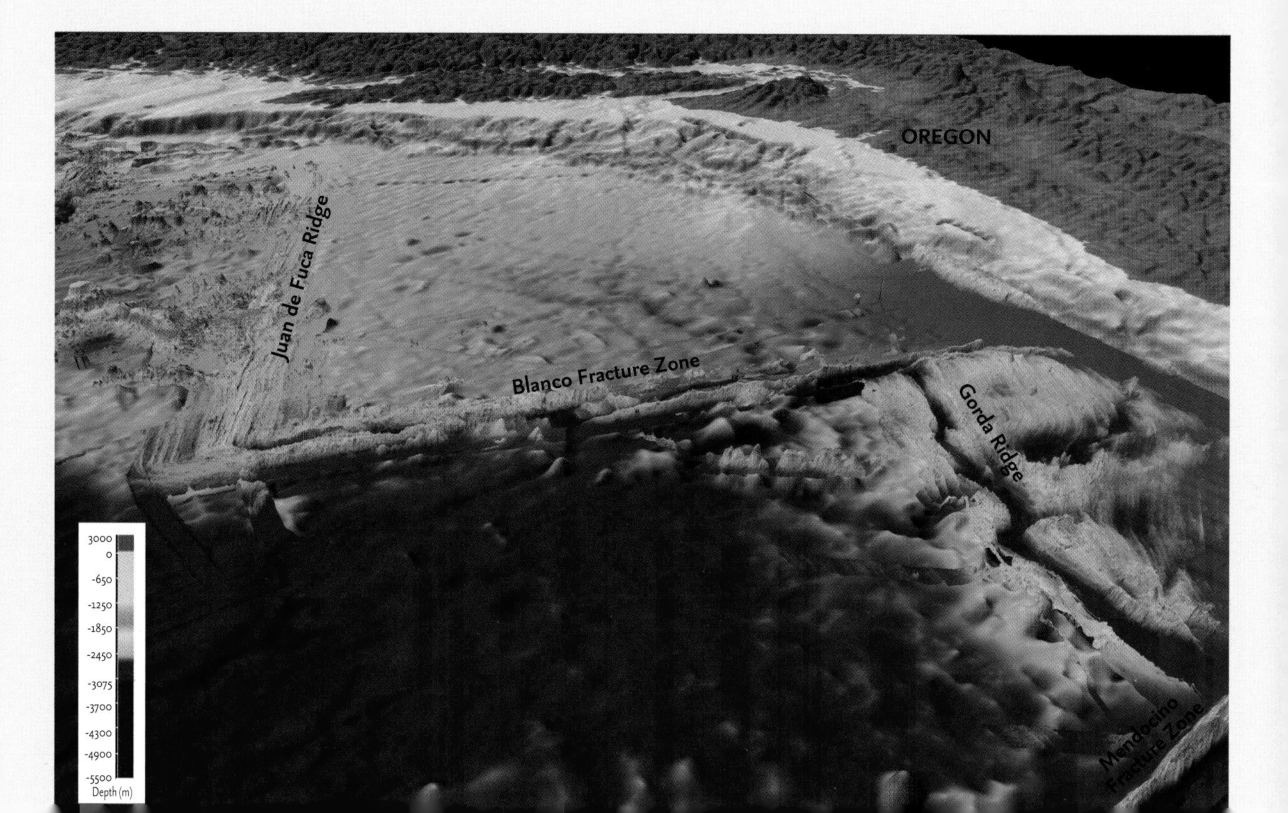

RIGHT: Perspective view of the northeast Pacific Ocean and adjacent North American continent showing the prominent ocean ridge between the Pacific plate (bottom) and the Juan de Fuca, North America, and Gorda plates (top and right).

LEFT:
A portion of the Mariana Arc, western Pacific Ocean viewed from the south looking north. The curvilinear submarine volcanic arc is featured mid-image, with the island arc to the east (right). The top right corner of the image depicts the area of the Mariana Trench.

LEFT:
Characteristic pillow lavas found in areas of new seafloor formation such as spreading ridges or active undersea volcanoes.

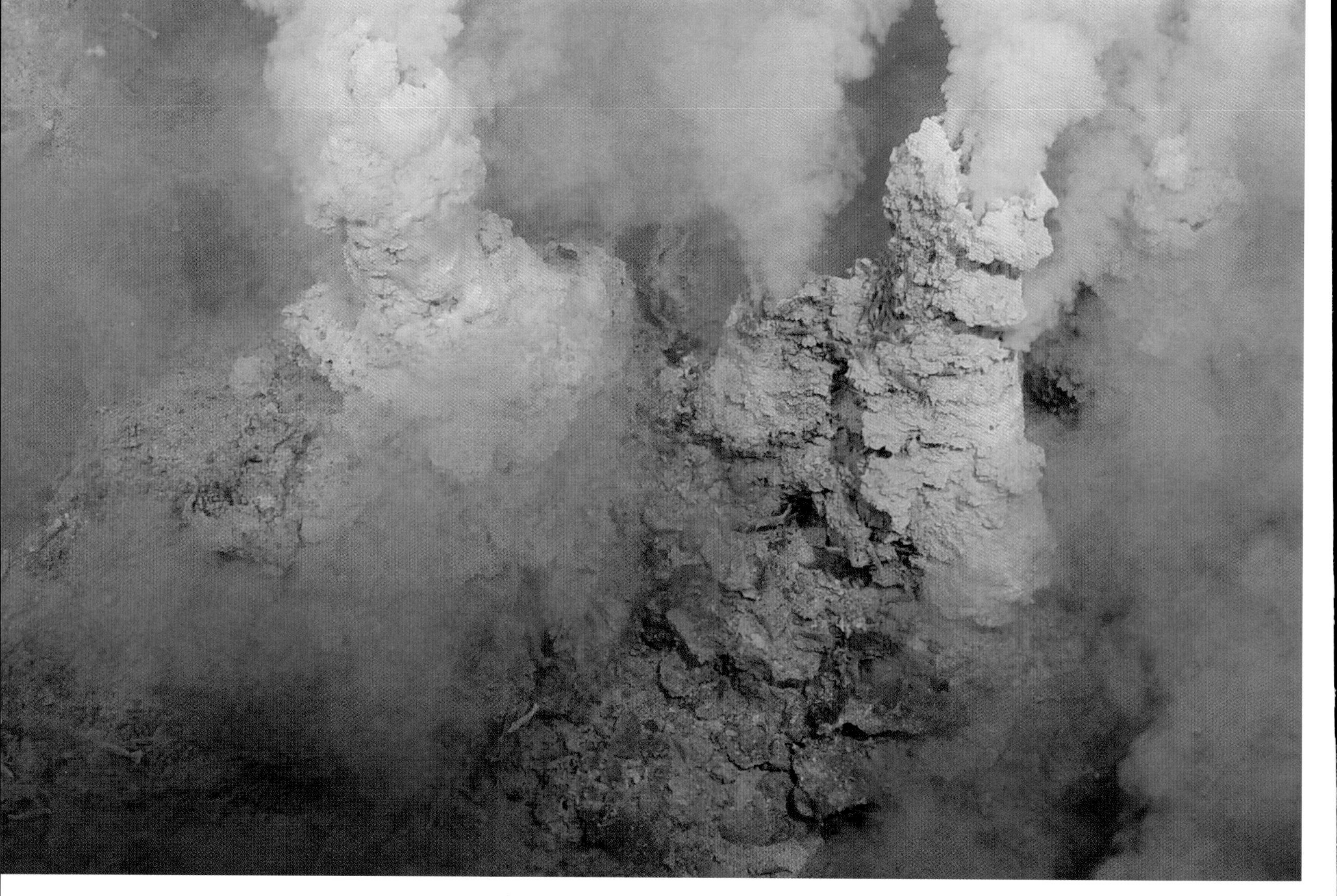

ABOVE:
White smoky vent fluid rising out of small chimneys at Northwest Eifuku volcano in the northern Mariana Arc, western Pacific Ocean. This area was named Champagne vent because bubbles of liquid CO_2 were rising out of the seafloor.

Hydrothermal vents and seeps

In the late 1970s a stunning discovery was made on the Mid Ocean Ridge (MOR) east of the Galapagos Islands. Researchers had for some time hypothesized that there should be hot springs on the ocean floor just as there are on the continents. Hot springs on the continents are found in volcanic areas (for example at Yellowstone) and result from the interaction of magma with the water table, resulting in boiling water and precipitation of sulfate minerals such as gypsum. The cold seawater sinking into cracks on the seafloor at the volcanically active crests of the MOR interacts with underlying magma, is heated up hundreds of degrees and leaches minerals from the surrounding rocks before rising back up through the seafloor to exit as warm or hot springs. The 1977 discovery of hydrothermal vents on the Galapagos Ridge was one of the most important discoveries of twentieth-century science. It altered our view of life on earth and opened up new possibilities for the very origin of life. An entire new ecosystem was discovered which is based not on photosynthesis, where the food chain is based on solar energy, but on chemosynthesis, where the food chain is based on chemical energy. Specialized microbes base their metabolism largely on using certain chemicals released at hydrothermal vents to fuel microbial biochemical reactions. These specialized microbes are the basis of ecosystems which depend on the presence of such chemicals. Bizarre animals such as giant tubeworms, clams, mussels, and specialized worms have evolved to be completely dependent on the chemical energy by incorporating symbiotic microbes into their tissues. Hundreds of seafloor vents have been discovered along virtually all sections of the MOR since 1977.

The high pressure from the overlying ocean allows temperatures as high as about 400° C (750° F), with the exact boiling temperature dependent on the depth. These fluids, laden with minerals and gases, precipitate iron, zinc, and copper minerals upon mixing with the near-freezing ocean. The jet of fluids emitted from these seafloor hot springs are called "black smokers." The long-term venting over decades builds up chimneys and mounds of minerals such as pyrite (iron sulfide), chalcopyrite (iron-copper sulfide) and sphalerite (zinc-iron sulfide).

There are other places where chemical-laden fluid exiting the seafloor feeds chemosynthetic life. These "seeps" occur in many different settings, from shallow sites on the continental shelf to the deep ocean trenches. These occur above large pockets of sub-seabed hydrocarbons like those in the Gulf of Mexico or where methane is being released from its ice-like form (clathrates) within large areas of continental margin and subduction zone sediments.

ABOVE:
Mussels at Northwest Eifuku volcano. In places they are so dense that they obscure the bottom. The mussels are about 7 in (18 cm) long and the white crabs are about 2.5 in (6 cm) long.

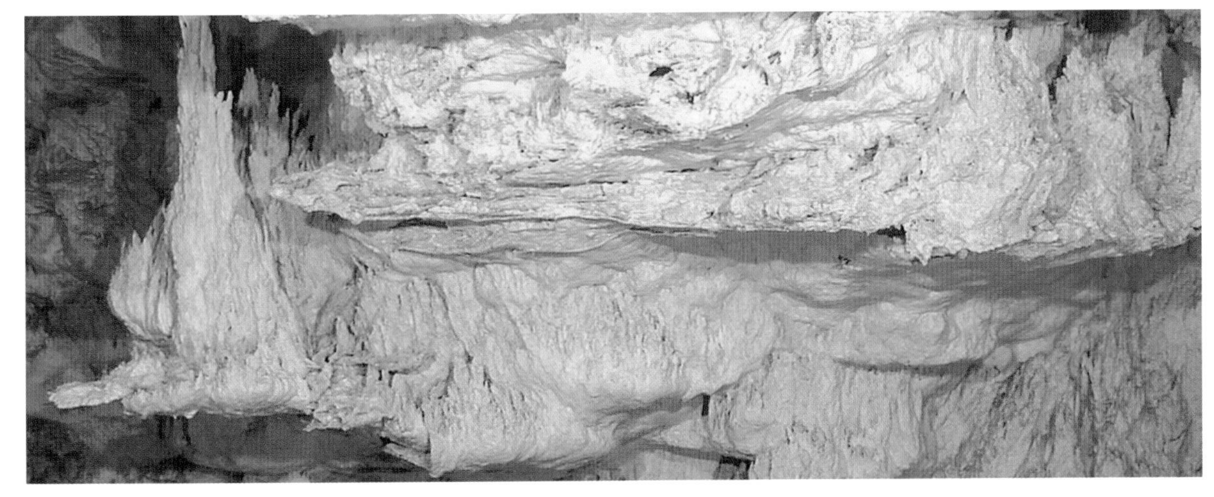

TOP LEFT:
This delicate flange actively vents heated hydrogen and methane rich fluids. These unique carbonate (limestone) structures are located on the IMAX Tower at the Lost City Vent Field off-axis from the Mid-Atlantic Ridge at 30 degrees North Latitude. Lost City is currently the only known hydrothermal field composed solely of carbonate chimneys.

BOTTOM LEFT:
Degassing event at Brimstone Pit, western Pacific Ocean, which started releasing an escalating number of bubbles (probably CO_2) as the plume cloud increased in volume. Pieces of sulfur can be seen at the base of the cloud.

Monitoring seafloor volcanic and hydrothermal activity

Most of the earth's volcanic activity is hidden from direct observations beneath a mile or more of seawater. Sunlight is absorbed below about 650 ft (200 m) in the oceans, so most of the deep oceans have to be explored using sound energy. The global network of seismometers on continents and islands can monitor larger submarine earthquakes caused by faulting along the plate boundaries, but it is not sensitive enough to monitor most submarine volcanic activity because the energy produced during volcanic events (mostly due to movement of magma) is much less than that produced by motion along the major faults. Fortunately, a network of fixed hydrophones installed in the ocean by the U.S. Navy in the 1960s became accessible to the scientific community in the early 1990s. This system is very sensitive to sounds transmitted directly through the ocean. Although only a small section of the mid-ocean ridge lying off the western U.S. and Canada can be effectively monitored with this system, several eruptions have been recorded in real time since the system became operational in 1993. In the follow-up expeditions to these sites, scientists have measured the volume of new lavas, probed giant plumes of warm fluids accompanying the eruption, and discovered new forms of life existing in the rich chemical broths generated by the magmatic heating of seawater.

There has also been some recent success in direct observation of undersea eruptions in the deep ocean. In 1996 NOAA (National Oceanic and Atmospheric Administration) scientists studying the Juan de Fuca Ridge off Oregon decided to place several monitoring instruments on the summit of Axial Volcano, an unusually shallow portion of the ridge, to attempt to catch the next eruption. This part of the ridge lies over a "hotspot"—an anomalously warm place in the underlying mantle where excess magma is produced—so it seemed to be the most likely location for a seafloor eruption. This marked the beginning of the New Millennium Observatory (NeMO), the world's first submarine volcano observatory.

In late January of 1998 Axial Volcano had a major volcanic eruption. Bottom pressure meters recorded the movement of the seafloor as magma migrated from under the caldera (large crater) during the eruption, a phenomenon commonly observed on continental volcanoes but never before in the deep ocean. A second pressure meter was captured by the lava flow and had to be pulled out of the lava the next year! In addition, temperature devices placed on vertical moorings recorded the large pulse of warm water caused by the eruption. In the years since the 1998 eruption, careful measurements have shown a steady rise of the caldera floor—very likely due to new magma slowly making its way up through the earth to reside in the shallow magma chamber beneath the caldera.

A lot was learned from the 1998 event (see http://www.pmel.noaa.gov/vents/nemo/index.html) but still lacked some very important information about mid-ocean ridge eruptions. The application of a new technology developed for a deepsea tsunami warning system offered a solution. In this system, called NeMONet, instruments on the seafloor transmit data through the water column to a surface buoy that relays it up to a satellite and then back to a shore laboratory. The NeMO site recently was chosen as one of the primary nodes on a planned seafloor cabled observatory called NEPTUNE, which should become operational in the next five years. (see http://www.neptune.washington.edu/pub/whats_neptune/whats_neptune.html).

RIGHT:
Eruptive underwater activity at Brimstone Pit in the western Pacific Ocean. The pressure of 1,837 ft (560 m) of water over the site reduces the power of the explosive bursts. Also, the water quickly slows down the rocks and ash which are violently thrown out of the vent.

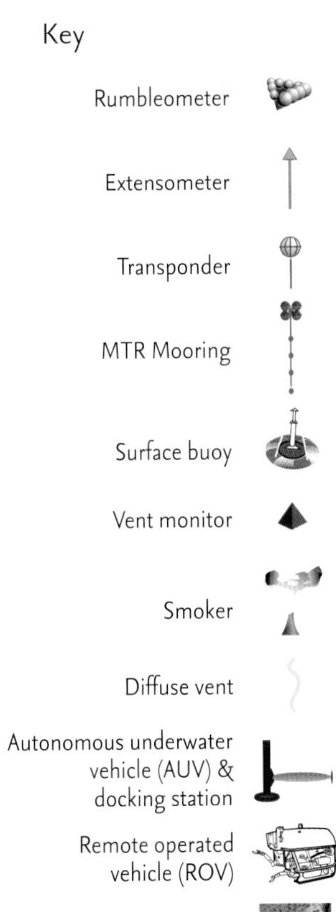

ABOVE:
Plan for NeMO (New Millennium Observatory) in the caldera of Axial seamount on the Juan de Fuca Ridge. The area of 1998 lava flow is shown in gray as well as the approximate positions of instruments prior to eruption in January, 1998.

BELOW:
Oblique view of Axial Volcano, showing the pattern of earthquakes on the summit and extending down the southern flank of the volcano.

RIGHT:
Black smoker at a mid-ocean ridge hydrothermal vent in the Atlantic Ocean.

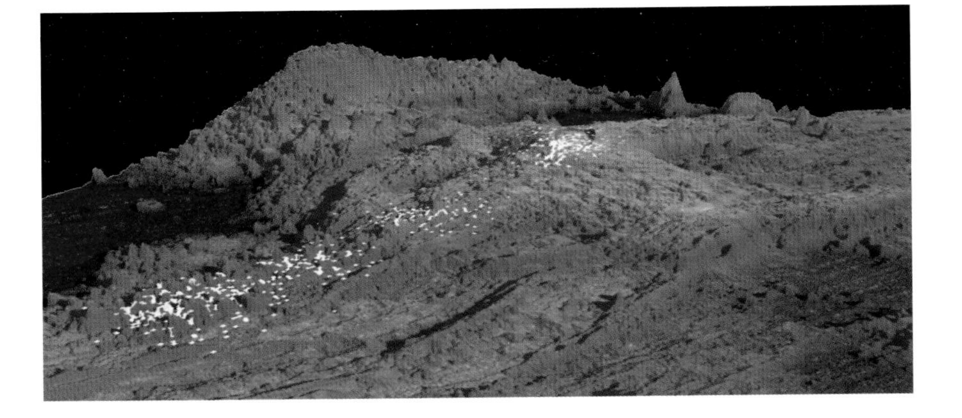

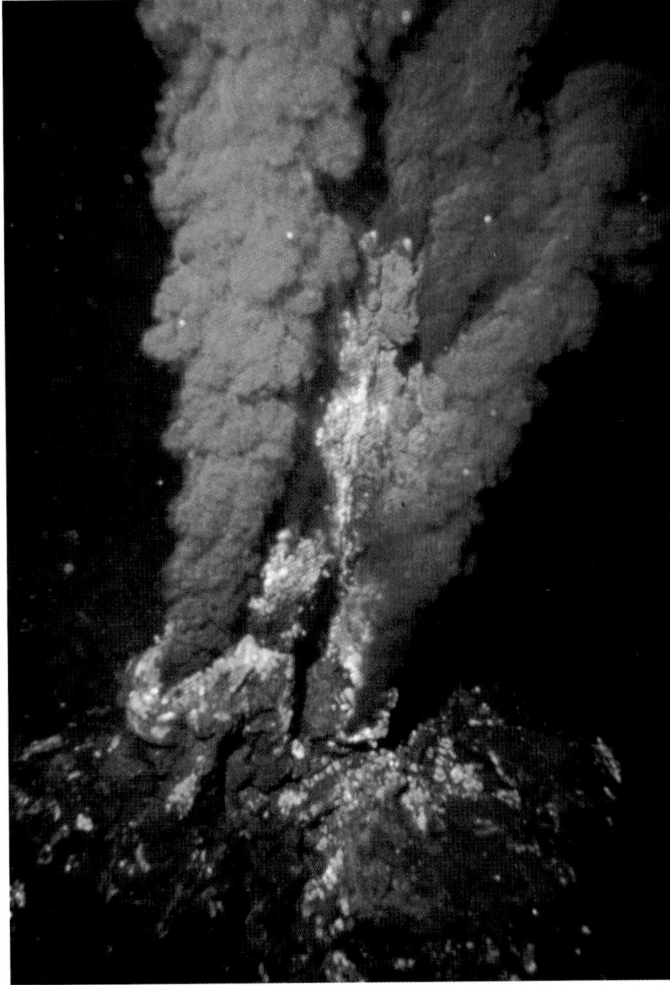

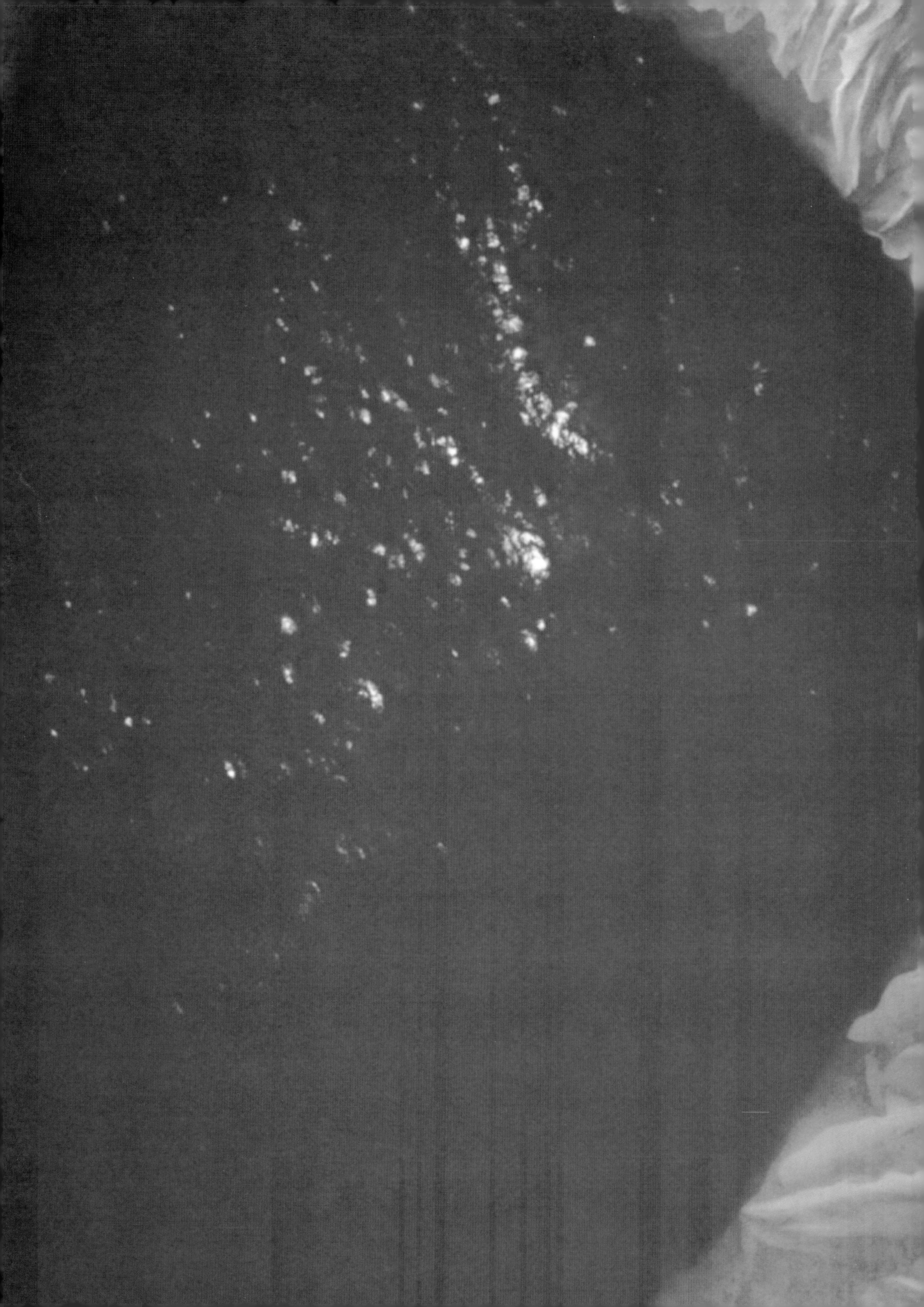

Ocean zones

Ocean zones
Introduction

Staring out at the ocean from land, or even from a ship, it is easy to have the impression that the ocean is a homogenous environment for as far as the eye can see. With a little traveling around one might recognize differences between the animals from the warm waters of the tropics and those from cooler waters away from the equatorial zones. Still, the perception of uniformity within broad ocean areas persists, even to a degree among specialists. Time after time the trend in studying the ocean has been to find more and more patterns of variation in plants and animals, water chemistry, and physical features which distinguish different zones of the oceans. Many of these patterns affect humans or the species we depend on for our survival. So let us dive in and begin to explore some of these patterns and current issues surrounding the zones of our oceans.

At the broadest level the oceans can be divided into zones based on distance from the coastline and bottom topography. Coastal waters are greatly influenced by the adjoining terrestrial environment and generally extend away from shore for only a few miles or so. Thus coastal waters are also generally shallow. A great diversity of ecosystem types from coral reefs to mangrove forests to rocky coasts occur in this zone. The chapter on vital ecosystems (page 134) discusses these coastal ecosystems in detail.

Beyond the coastal waters is a moderately shallow area with depths of less than about 400 ft (122 m) and a gentle gradient of around 1 in 500 called the continental shelf. This shelf averages about 40 miles (64 km) in width but in some areas can be up to 400 miles (644 km) wide. The edge of the continental shelf is generally distinct and leads to a steep-gradient (1 in 20 or so) zone called the continental slope which extends down to the ocean floor at around 13,000 ft (3,960 m). This slope zone can be bisected by valleys called submarine canyons.

The ocean floor or deep-sea bottom is a flatter zone with depths generally from 10,000–20,000 ft (3,050–6,100 m; average ocean depth is 13,000 ft/3,970 m). It is not a featureless place however. The deep-sea bottom contains vast, towering mountain ranges, individual seamounts, hills, oceanic ridges, deep trenches, and a host of other topographic and biotic features. In areas where hot or cold liquids and chemicals shoot from the ocean floor, unique hydrothermal vent or cold-seep communities occur which support bizarre organisms found nowhere else. These features all affect ocean circulation and ecosystems.

PREVIOUS PAGE: Great Bahama Canyon from space. This is part of a branch of the Canyon which is called the Tongue of the Ocean. The dark blue waters of this abyss contrast with the light blue waters of the shallow Great Bahama Bank, marking a depth differential of hundreds of feet. The Canyon, over 140 miles (225 km) in length, reaches a width of 23 miles (37 km) and a depth of 2.5 miles (4 km).

BELOW: Shallow reef in Guam, Mariana Islands

BELOW: Tube sponge on the McGrail Bank, Gulf of Mexico.

RIGHT:
Alfonsino, a relative of coral reef squirrel fishes, hovering around a large Lophelia coral on the North Carolina continental slope.

We can describe zones of the ocean in another way, based on depth and relationship to the ocean floor. The pelagic zone is the open water areas of the ocean. This zone is obviously dominated by factors related to water chemistry and physics, and is affected to a much lesser degree by interactions with the ocean floor. In contrast, the benthic zone is the zone near to the ocean floor. The physical topography and surface materials and processes play a much more direct and important role in what occurs in this zone. The demarcation between these zones is less precise because the balance of the influence between water and ocean floor varies with distance from the ocean floor or coast in ways which depend on the specific animals or chemical or physical processes involved. However, many organisms are broadly specialized to live in one or the other of these zones.

The remainder of this chapter considers in detail some of the exciting new knowledge and issues regarding some of these zones.

RIGHT:
High-definition colored sonar image of the continental shelf and slope off the central California coast, U.S.A. This was made by ships recording sound echos from the ocean bottom. Colors show depths of the ocean floor, from white (near sea level), through orange, yellow, and green, to blue – a depth of 9,840 ft (3,000 m).

RIGHT:
Hermit crab from a cold seep site near New Zealand.

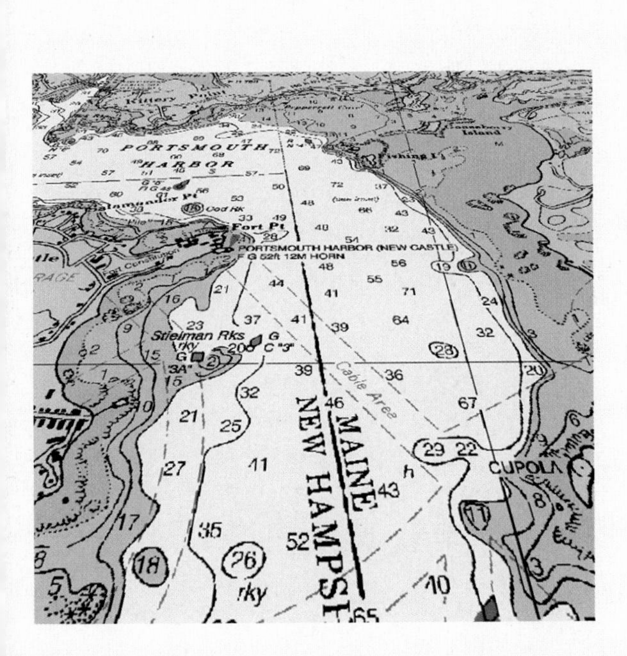

LEFT AND RIGHT:
A classical style chart of Portsmouth Harbor, New Hampshire, (left) gives little indication of the underlying bottom configuration of this portion of the coastal waters benthic zone. However, the underlying data obtained by a modern multi-beam sounding system (right) can produce three-dimensional views showing rock outcrops, deep holes, and sediment-covered areas, which help scientists better understand the environment of the area and plan scientific studies.

Ocean subzones

The pelagic and benthic zones can be divided into a number of functional subzones which reflect correlated patterns of physical features and biotic communities. The benthic zone can be divided into a deep-sea bottom benthic subzone and corresponding subzones in the continental shelf, continental slope, and coastal waters zones. Many commercially important food resources occur in the continental shelf benthic zone including lobsters, shrimp, scallops, and flatfishes like halibut and flounder. Most commercial exploitation for mineral and hydrocarbon fuels occurs in this zone as well, although recent advances in drilling technology mean just over 50 per cent of active petroleum leases in the Gulf of Mexico are in depths of over 1,000 ft (300 m).

The pelagic zone can be divided into three main subzones as well: epipelagic, mesopelagic, and bathypelagic. The epipelagic zone extends from the surface down to about 700 ft (210 m) depth, beyond which there is rarely any significant light. This zone contains the vast majority of commercial fishes and is home to many protected marine mammals and sea turtles. Many key ideas and human impacts discussed elsewhere in the book mostly affect this subzone.

The meso–(middle)–pelagic subzone extends from 700 ft (210 m) to about 3,300 ft (1,000 m) depth. This is a zone where light, temperature, and water chemistry transition from surface zone conditions to their relatively uniform state in the deeper subzone. Many organisms within the mesopelagic zone undergo daily vertical migrations into the upper ranges of this zone during the night and back down during the day. Many epipelagic fishes and marine mammals feed on these migrating organisms.

The bathypelagic zone is in the deeper ocean waters below about 3,300 ft (1,000 m). This zone is uniformly cold (about 40°F/4°C) and dark. Food is scarce, mostly coming from decay and fecal matter of organisms living in the upper subzones. Organisms have many unusual adaptations for living in this subzone.

OPPOSITE PAGE: Spinner dolphins at Sharm El Sheikh, Egypt.

OPPOSITE PAGE: Green sea turtle at the Great Barrier Reef, Queensland, Australia.

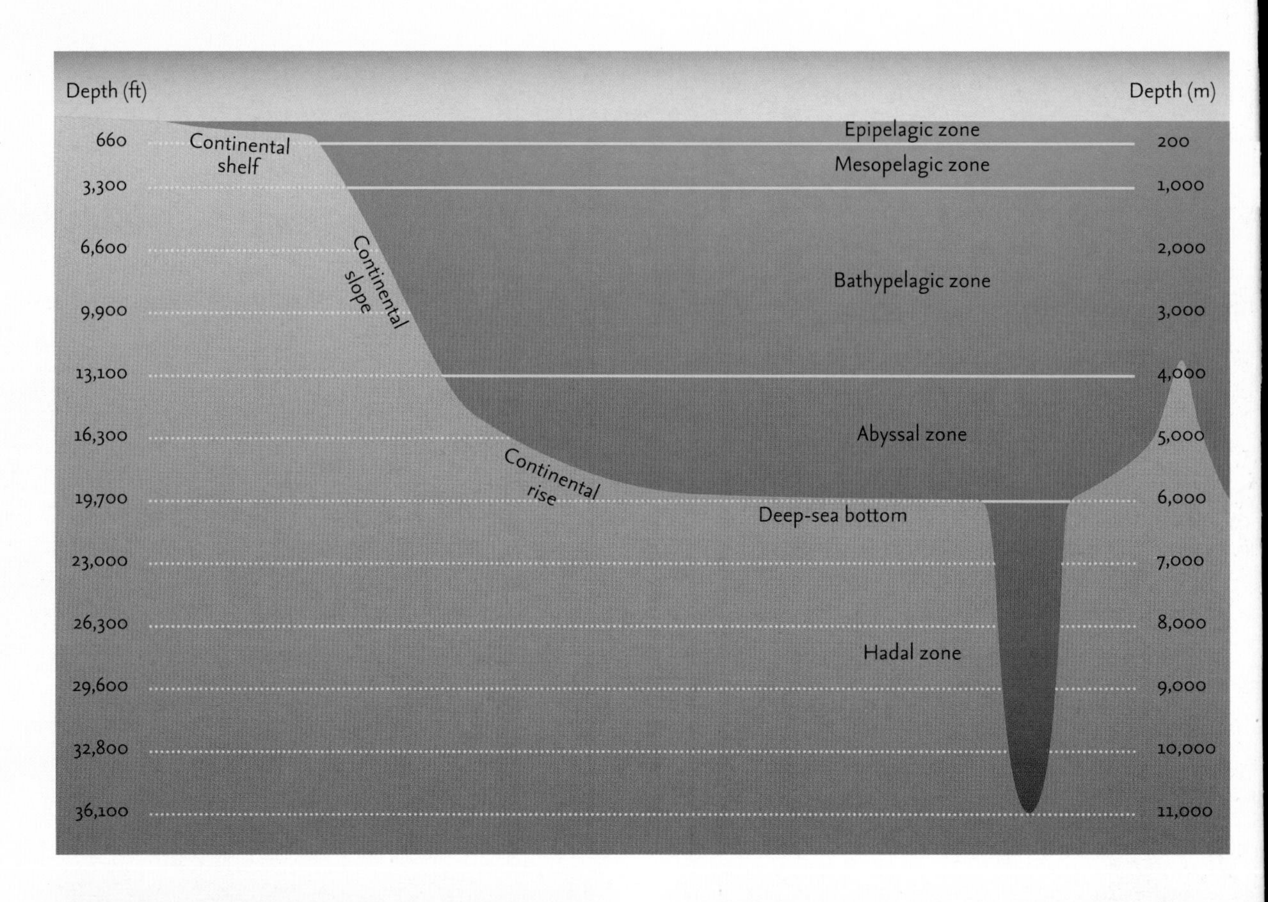

RIGHT: Zones and subzones of the ocean.

RIGHT: Benthic octopus and clam at a depth of 4,800 ft (1,460 m) on the Davidson Seamount, California.

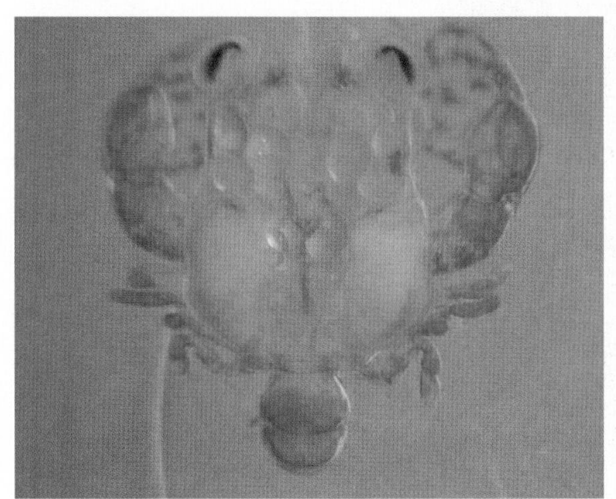

FAR RIGHT: Crab larva, a form of zooplankton found in the pelagic zone. This specimen was found southeast of Charleston, South Carolina.

Large marine ecosystems

Large areas of the oceans function as ecosystems, with natural factors, along with pollution from air, land, and water, and overexploitation of living resources, influencing their varying productivity. Large marine ecosystems (LMEs) are distinctive ocean regions, 77,200 square miles (200,000 sq km) or greater, which encompass coastal areas from river basins and estuaries out seaward to the break or slope of the continental shelf, or out to the seaward extent of a well-defined current system. Each LME is characterized by a distinct and unique bathymetry (depth), hydrography (tides, currents, upwellings, and physical conditions of ocean waters), and biological productivity whose plant and animal populations are inextricably linked to one another in a food chain.

Ecologists and resource managers attempt to understand the various interactions and the major driving forces which cause change in the inner structure of the LME and variability in LME productivity. They consider these when making resource use and management decisions.

Deep water corals

TOP LEFT:
Thorny sea star and small pink coral at 4,450 ft (1,356 m) on the Davidson Seamount, California.

TOP RIGHT:
Black coral bush in the Flower Garden Banks National Marine Sanctuary, Gulf of Mexico.

RIGHT:
Mushroom soft coral with polyps extended at 4,820 ft (1,470 m). It has the ability to push water away with a "pulsing" motion.

BELOW:
Bubblegum coral.

When most of us think of corals we think of warm waters, pretty colors, sand, and sun. But in recent years scientists have learned about the widespread occurrence and ecological importance of another group of dozens of species of corals which live in the benthic zone of the outer continental shelf and continental slope subzones from depths of 100 to over 6,000 ft (30–1,830 m). These deep corals do not have the kind of symbiotic algae which their shallow-water cousins use to take advantage of sunlight. Instead they mostly feed on planktonic animals. Deep corals are often bushy or branching and serve as home to hundreds of other species of invertebrates and fishes which live on or within their structure. Some live for hundreds of years.

Some deep-water corals like black corals and pink corals are commercially harvested for jewelry and herbal medicines. Black and pink corals are currently regulated by the UN Convention on International Trade in Endangered Species Appendix II, which means that they may become endangered if trade is not controlled.

The major threat to deep corals is destruction of their habitat from fishing trawls and other gear which inadvertently entangles and rips them up. Oil and gas exploration and production and the laying of pipelines can also destroy these amazing creatures. Recently, scientists have realized that acidification of the oceans from increased human-generated carbon dioxide input may harm these animals.

The National Ocean and Atmospheric Administration (NOAA) has recently instituted protections for deep corals in many U.S. Fishery Management Plans and is developing a Deep Coral and Sponge Research, Conservation and Management Strategy. The ivory bush coral is a Species of Concern, which means voluntary protections are being encouraged in order to prevent it from being put on the endangered species list.

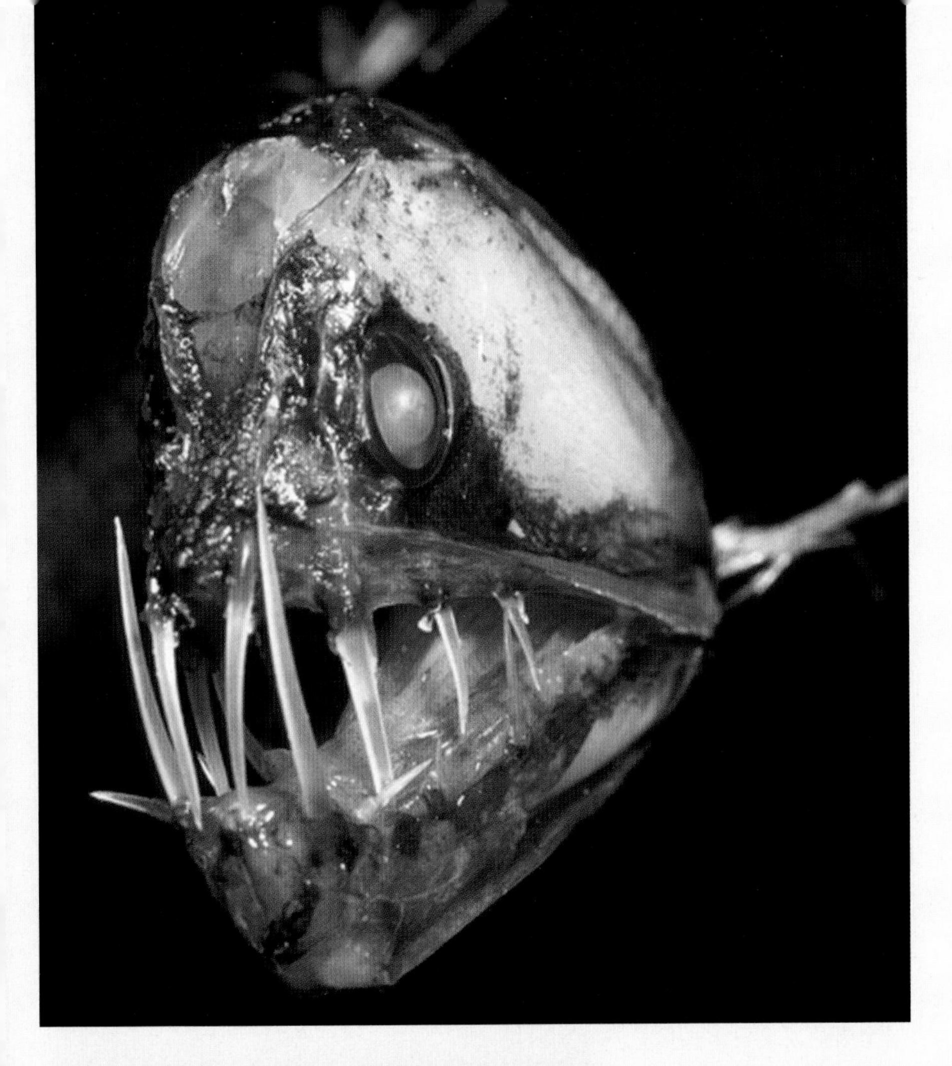

ABOVE:
Viperfish, a deep-sea fish with long, needle-like teeth and hinged lower jaws. It grows to lengths of 12–24 inches (30–60 cm).

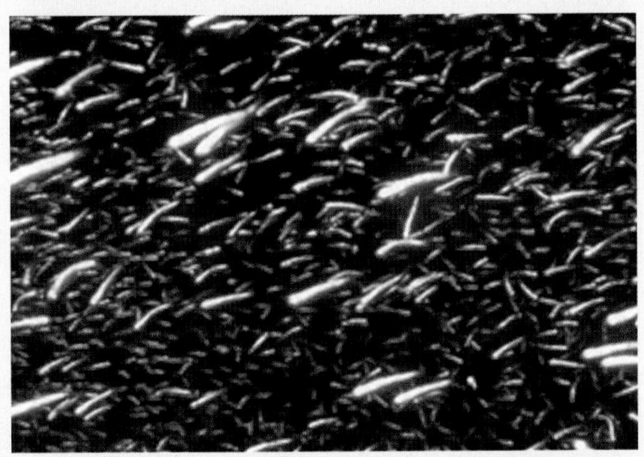

LEFT:
Dense nighttime aggregation of light organ fishes using bioluminescent photophores as seen from below in water off Oahu, Hawaii. During daytime these fishes may be camouflaged by producing this light while at night the many light sources may confuse predators.

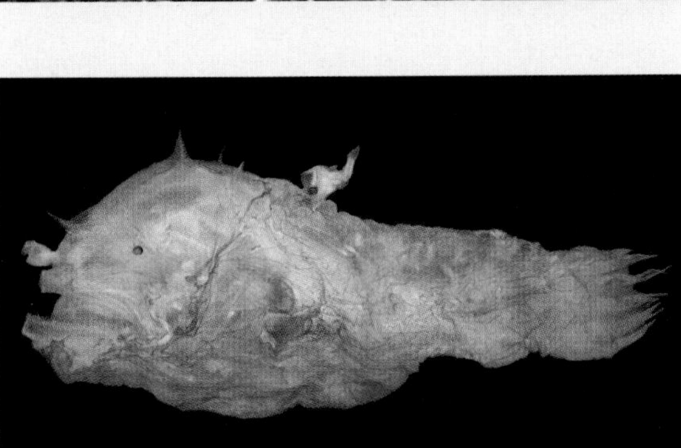

LEFT:
Deepwater anglerfish female with single parasitic male, measuring only 0.24 inches (6.2 mm), attached to its back.

Deep-sea fishes

Deep-sea fishes are an important energy link in ocean ecosystems; they serve as food for predatory animals in the deep sea as well as those in the epipelagic subzone. For example, many large whales and dolphins are now known to feed on the large groups of fishes which migrate vertically towards the surface every night. These groups are often called the "deep scattering layer" because of the dense sonar returns seen when sound is bounced off their swimbladders. Recent research has shown that some dolphins even follow the horizontal migrations of these fishes during the course of the night.

Deep-sea fishes have evolved unique adaptations to deal with the physical and ecological features of their environment. Mesopelagic fishes live in a subzone with small amounts of light. These fishes vertically migrate, and in order to find their food in this "twilight" zone, many have evolved eyes which can see under very low light conditions. Some of these special eyes consist of long "tubular" eyes which focus and concentrate light; others have highly reflective retinas and other structures to maximize vision. To camouflage themselves from the eyes of other predators, many mesopelagic fishes have evolved bioluminescent light organs (called photophores) which use symbiotic bacteria or chemical light generation similar to that of a firefly in order to obscure their silhouette, particularly from below.

Bathypelagic fishes live in the dark zone lacking sunlight, so they tend to have reduced eyes and rely on other senses. As food is scarcer in this zone too, these fishes often have large mouths, extensible stomachs, and reduced skeletal structure. Because of the large volume of water and low fish density, bathypelagic fishes tend to have unusual reproduction strategies, including males which become parasitic "sperm sacks" which permanently attach to females in some species of anglerfishes.

Water and seawater

Water and seawater
Introduction

The ocean represents many things to many people. Some see it as a large recreational area, others as the source of their income, and some even believe it to have a mystical presence on the earth. Whatever their opinion or connection, the foundation of the ocean is water. Before the pyramids, before paintings on the walls of caves, and before the discovery of fire, every part of the globe was connected because of the constant moving and mixing of the water.

This connection by water does not follow any boundary arbitrarily drawn to divide nations; it reaches across them and binds cultures. The ocean is massive and no matter what, regardless of the current events of the time, the ocean continues to churn its water and travel around the world.

The seawater is changing and scientists are working to understand the significance of the effects. The ocean is not like land where changes are readily apparent to the casual observer. Forests burn and glaciers melt but the ocean suffers change far away from even the most dedicated witness. Scientists must spend large amounts of money, travel dangerously far from civilization, and use complex equipment to sample the many different properties of seawater. Still, after overcoming all these obstacles, the most successful research cruise can only cover a tiny percentage of the vast ocean.

Local problems with seawater can quickly become problems far away from the source. Increased use of fertilizers can cause massive hypoxic zones 1,000 miles downstream. A hypoxic zone is an area with low levels of dissolved oxygen. The low oxygen levels suffocate bottom-dwelling organisms which cannot swim or move out of the area. These "dead zones" then set off a chain reaction of negative impacts up and down the food chain which radiate across seas.

An oil spill may drift and travel on currents thousands of miles from the accident. This journey leaves a path of destruction with effects which last for many years after the initial incident. Even though the ocean does not appear different after a serious incident, it is never the same again.

Public awareness of this global tragedy is increasing. Attention from the media and inquiries by governments are bringing the changing ocean to the forefront of public consciousness. Scientists are working with industry to develop tools such as satellites and Autonomous Underwater Vehicles, which allow for the wide-scale study of the ocean. Most importantly, people are starting to view the ocean as a series of systems which are interconnected.

Seawater is the common thread which runs throughout all studies of the ocean. Geologists tracking underwater volcanoes must rely on chemists studying the substances mixed in the seawater. Meteorologists predicting the weather work with engineers who design satellites to observe the ocean from space. Even city planners developing the infrastructure of coastal cities depend on oceanographers to determine the cycling of elements through different compounds in seawater.

Everyone on the planet has a personal stake in this cooperation among scientists working to understand and improve the seawater. Scientists consider the ocean to be the origin of life on land. Remarkably, the chemical makeup of human blood is very similar to seawater. When animals crawled out of the ocean and started breathing air, they brought the seawater with them. As people travel around and determine the future of the Earth, their connection to the ocean courses through their veins. We are seawater. This tie will forever bind the future of humanity with the health of the ocean.

RIGHT:
Weather systems form over the Pacific Ocean and move across North America.

PREVIOUS PAGE:
Satellite images of Elands Bay, South Africa acquired on 2 February, 2002 (left) and 18 February, 2002 (right) showing the shoreward migration of an algal bloom or "red tide." Although some red tides form a healthy part of phytoplankton production, recurrent harmful blooms also occur. At Elands Bay about 1,000 tons of rock lobsters beached themselves during February 2002, when the decay of dense blooms of phytoplankton caused a nearshore hypoxic zone. The lobsters moved toward the breaking surf in search of oxygen, but were stranded by the retreating tide.

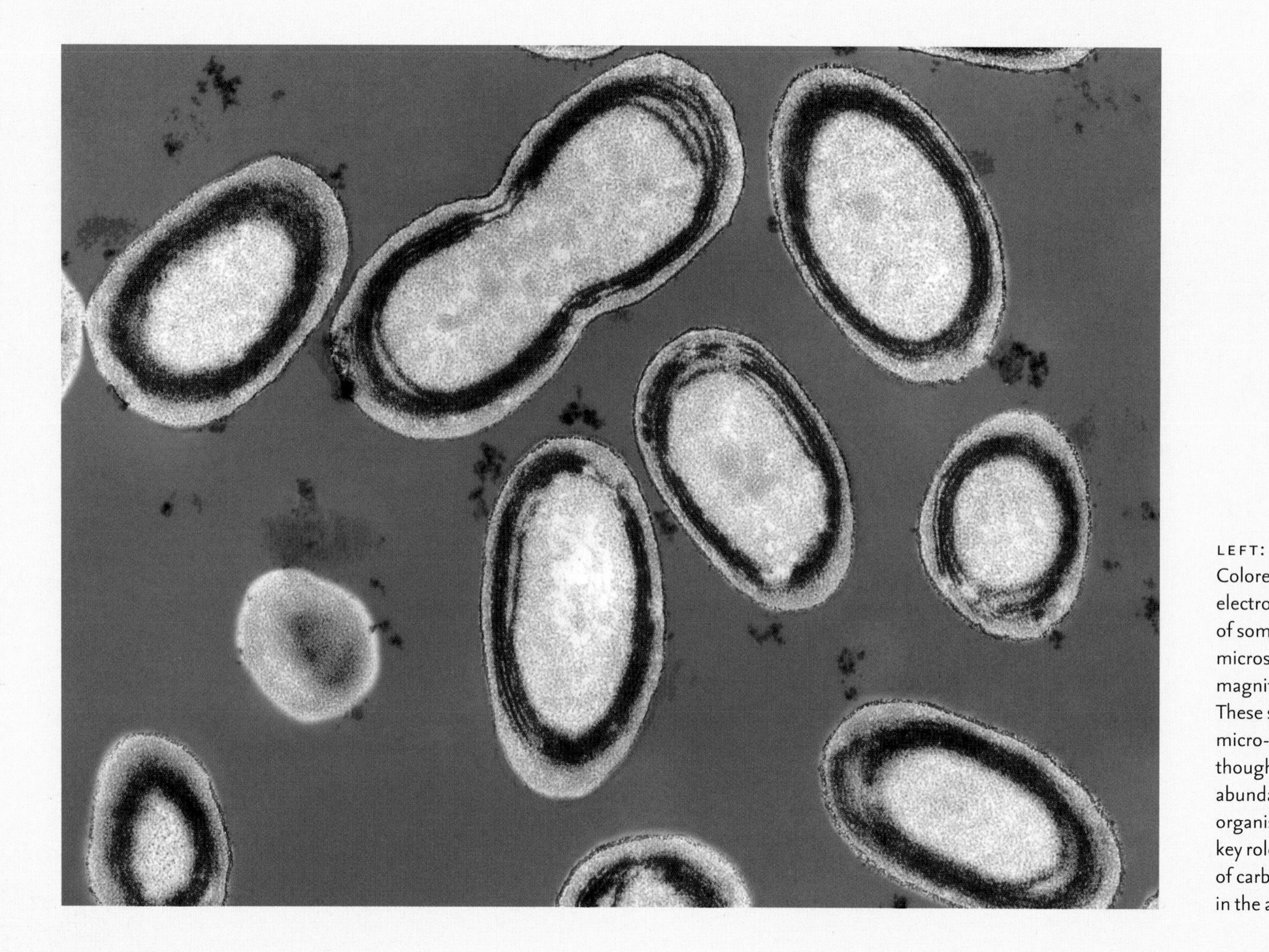

LEFT:
Colored transmission electron micrograph (TEM) of some photosynthetic microscopic plankton magnified 23,000 times. These single-celled marine micro-organisms are thought to be the most abundant photosynthetic organisms on earth, playing a key role in regulating levels of carbon dioxide and oxygen in the atmosphere.

Biogeochemistry

All animals and plants affect their surroundings through behaviors which involve chemical reactions. Biological activities such as digestion, respiration, and decomposition involve steps which create and destroy elemental compounds. The growing field of oceanic biogeochemistry attempts to understand the relationship between marine life and the processes which affect the cycling of elements through the ocean.

In 1934, Alfred Redfield was the first to view the composition of seawater as the result of many linked biological and chemical processes. He recognized that the ratio of carbon, nitrogen, and phosphorus in the bodies of marine plankton is the same ratio of the elements in the seawater which surrounds them. Redfield concluded that the elemental composition of the seawater would be different if the plankton were not using nutrients for biological processes such as eating, breathing, and reproduction.

Nutrients are chemicals necessary for biological processes. A limiting nutrient is a chemical which is necessary for growth, but is not available even though all the other necessary chemicals are on hand. The most common limiting nutrient in the ocean is nitrogen, although phosphorus, silicon, and iron may limit biological growth in some areas.

To understand limiting nutrients, scientists study the predictable and quantifiable patterns which cycle elements through nature. Only recently scientists have developed the sophisticated tools necessary to measure each step in the process. They measure the input and output of each step in the cycle and see if the quantities agree. The term "source" refers to a step in the cycle which provides most of the compounds for the next step in the cycle. For example, respiration by plants changes carbon dioxide into oxygen which animals then use for their own respiration. A "sink" does not allow a reaction's entire product to go to the next step. For example, an animal may store some of the product of a chemical reaction in the form of tissue or it may use it for reproduction. Sometimes the product forms a solid and falls to the bottom of the ocean. Regardless of the sink, the element will eventually re-enter the cycle at a different time and location.

ABOVE:
Scientists studying the north Atlantic right whale deploy a conductivity, temperature, and depth sensor (CTD) in the Great South Channel off Georges Bank. In addition to water density, the instrument also counts particles, photographs the sea floor, and measures dissolved oxygen.

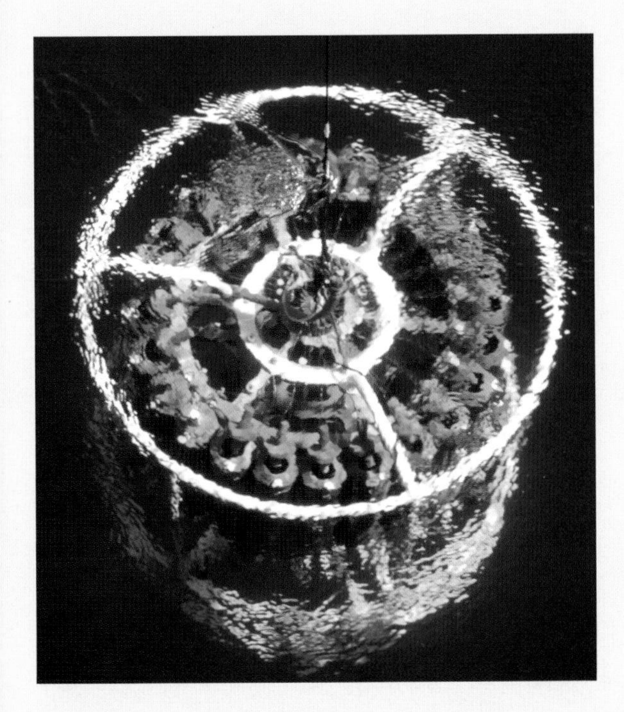

ABOVE:
Niskin bottles attached to a rosette are used to sample the ocean at discrete depths. When the rosette is at a desired depth, scientists aboard the ship press a button which closes a bottle. The rosette is then lowered to the next sampling depth.

ABOVE:
This plankton net is being deployed to collect near-surface plankton in Maug caldera in the Mariana Arc region, western Pacific Ocean as part of the Pacific Ring of Fire Expedition, April 2004.

Acidity and salinity

Water, in rivers and flowing over land dissolves compounds in the soil and washes them into the sea. Over millions of years, these compounds have created a chemical solution which is well mixed due to continual agitation of the ocean by currents and weather. Every element is represented in the seawater solution, although most are at minuscule concentrations. The most common types of compounds in the ocean are salts, due to the ease with which they dissolve in water.

Scientists call the amount of salt in seawater the "salinity." In some areas, the ocean acts as an evaporative basin and the water is very salty. An example of this happening is the salty Mediterranean Sea where the amount of water which evaporates every year is greater than the amount of rainfall and river runoff. Other areas have low salinity because of their proximity to large sources of freshwater. The Amazon River lowers the salinity of water more than 200 miles away from where it flows into the Atlantic Ocean.

Aside from chemicals entering the ocean from river runoff, surface seawater also absorbs chemical compounds in relation to their atmospheric abundance. For example, nitrogen is the most common element in the atmosphere and it is the most common dissolved gas in seawater. Some gases react with other chemicals when they are absorbed through the ocean surface and the result is an acid or a base. The measure of the acidity of a liquid is termed the "pH" level of that liquid.

A buffer is a chemical compound which allows a solution to resist changes in pH with the addition of an acid or a base. Freshwater lakes have no buffer system, which is why they are susceptible to acid rain. The ocean uses carbonate compounds as a buffer to keep the pH level from changing too rapidly in response to transient events. The increase in atmospheric carbon dioxide from burning fossil fuels, erupting volcanoes, and blazing forest fires has increased the amount of carbon dioxide dissolved in the ocean. Carbon dioxide forms carbonic acid when it dissolves in water. Computer models and laboratory experiments suggest that increased levels of carbonic acid may overwhelm the natural buffer system and lower the ocean's pH level. Scientists are currently studying whether corals and other organisms which depend on carbonate for their skeletons will survive if the pH is further lowered.

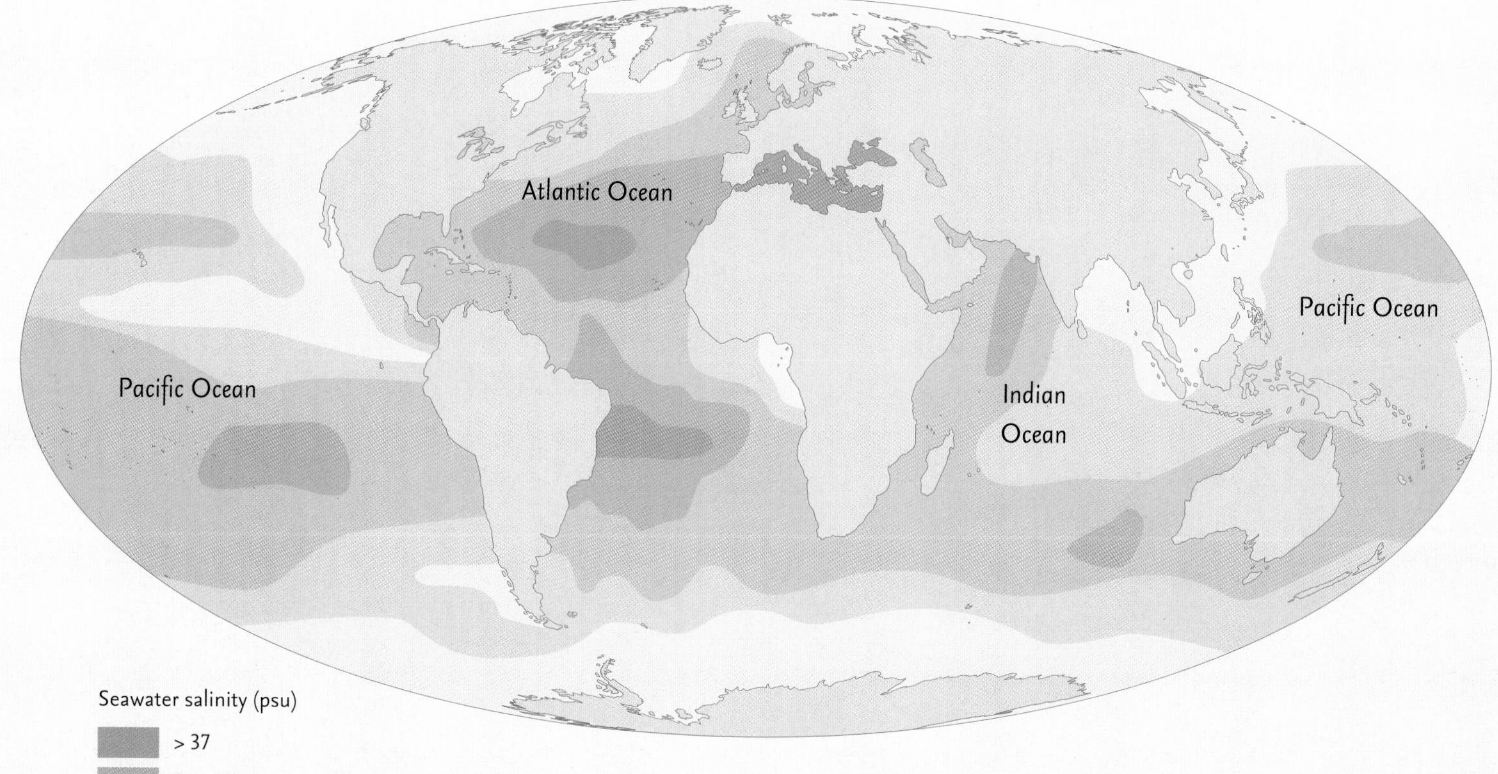

Seawater salinity (psu)

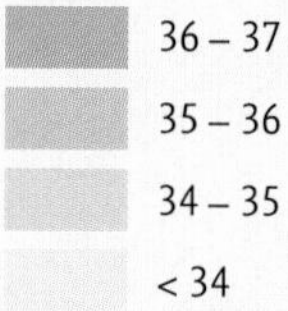

ABOVE:
Average annual salinity of the world's oceans. Highest salinities are typically found far away from the influence of rivers and major ocean currents. Salinity used to be measured as the amount of grams of salt dissolved per liter of water and expressed as parts per thousand with the symbol ‰. Modern instruments measure salinity as a ratio between the conductivity of a seawater sample and the conductivity of a standard potassium chloride (KCl) solution. As a ratio, salinity has no unit but many people use "psu" or practical salinity units to denote that an expressed number is a salinity measurement.

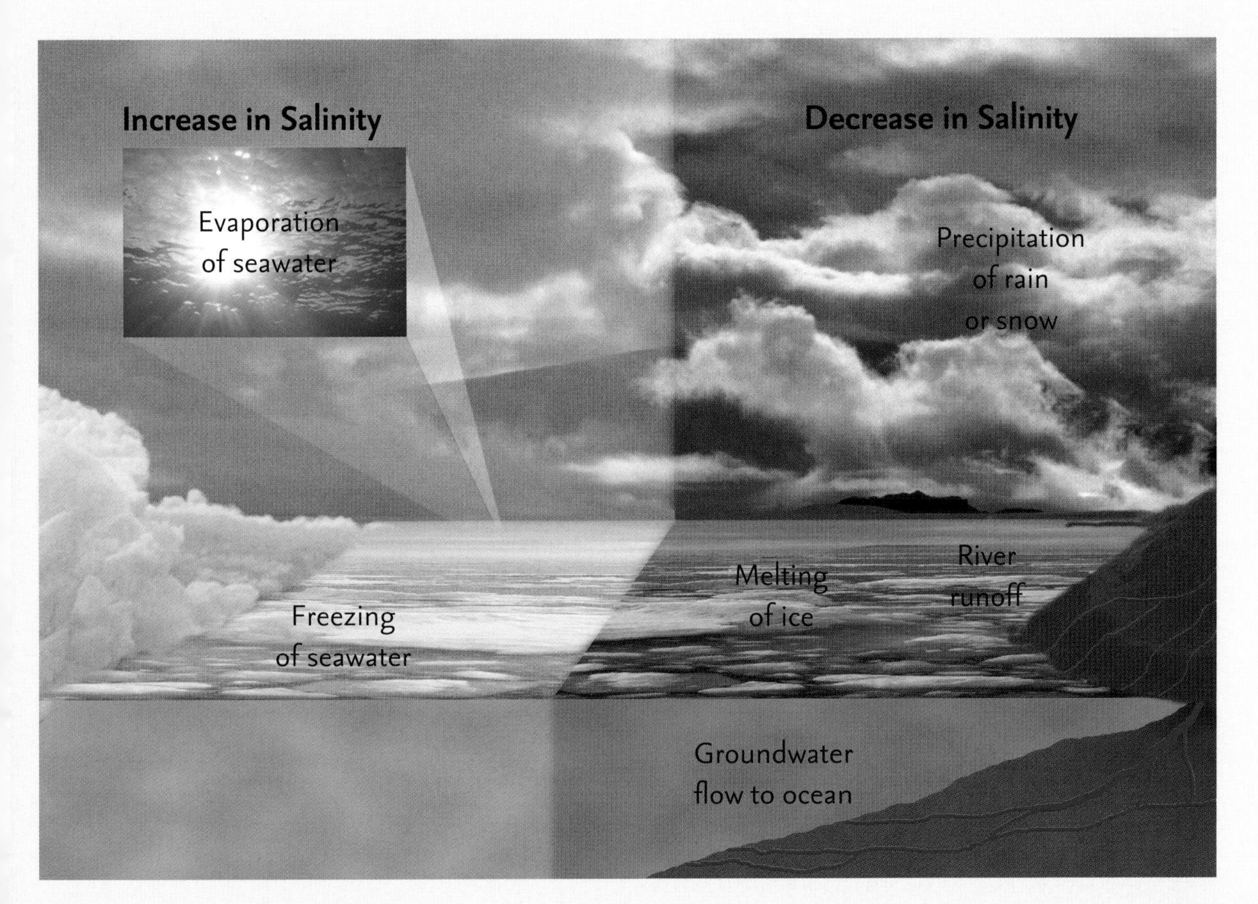

LEFT:
Processes which increase a body of water's salinity include evaporation and freezing. Dilution of seawater by precipitation, ice melting, groundwater flow, and river runoff decreases the body of water's salinity.

LEFT:
Close to the city of Manaus, Brazil the Rio Solimoes and the Rio Negro converge to form the Amazon River. The pale, murky color of the Rio Solimões heralds its burden of glacial silt and sand, which results from its origin in the Peruvian Andes mountains. The dark color of the Rio Negro is characteristic of clear waters which originate in areas of basement rock and carry little sediment. East of Manaus the pale and dark waters flow side-by-side as distinct flows before they eventually merge.

The hydrologic cycle

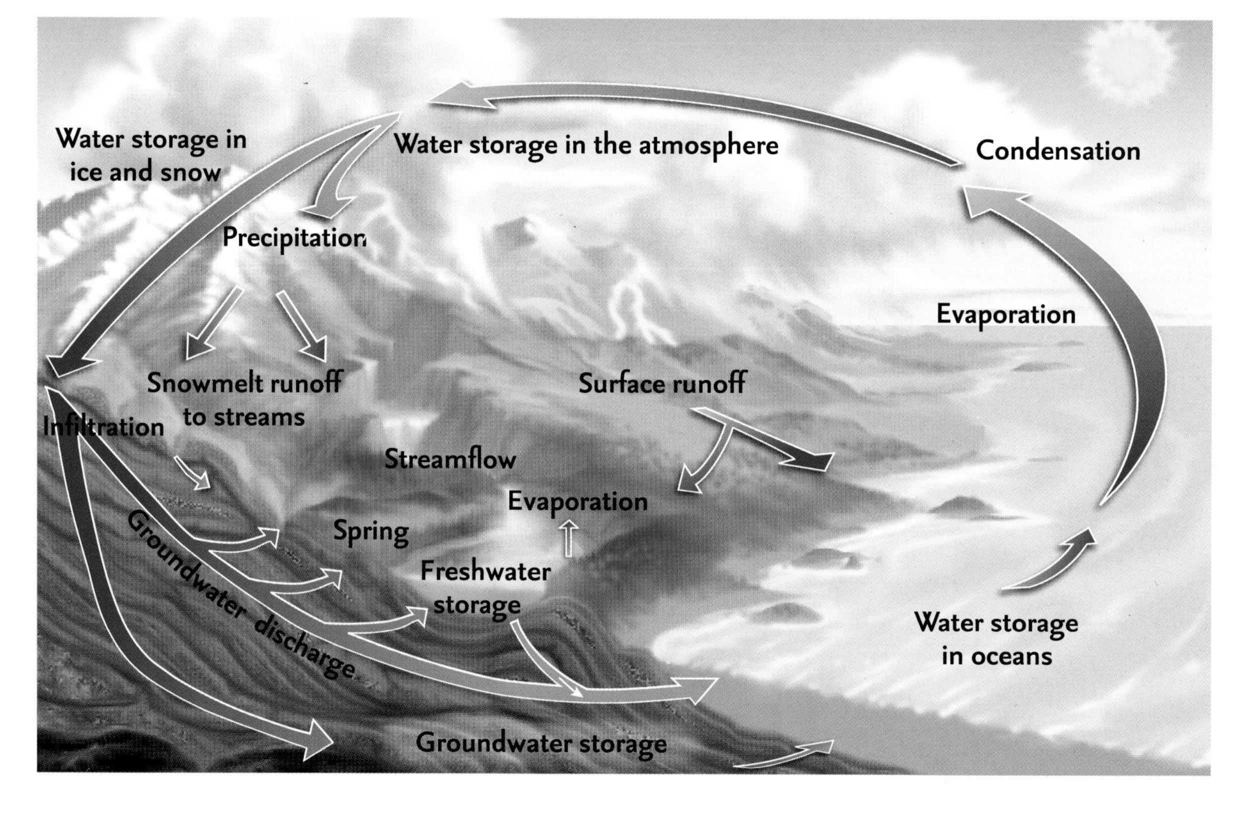

RIGHT:
The hydrologic cycle.

ABOVE:
Cumulus cloud formations over Palau, north Pacific Ocean.

The hydrologic cycle describes the movement of water around the globe as it changes states between solid, liquid, and gas. It is called a cycle because there is no beginning and no end, only a continuous loop of travel and transformation which has been occurring for millions of years. The drop of water coming out of a faucet today may have been a snowflake during the time of the dinosaurs, then part of the Mediterranean Sea when the ancient Phoenicians were beginning to learn navigation, and then inside an apple eaten by a pioneer traveling west during the California gold rush, all before washing into the city reservoir.

The daily passage of the sun heats up and evaporates water from the ocean. Because the ocean contains almost 97 per cent of the world's water, there is a large amount of water vapor put into the atmosphere every day. This water vapor traps heat from the sun which prevents the earth's surface from becoming too hot for anything to survive. Also, evaporation cools the earth by moving heat from the land to the atmosphere.

In time, the evaporated water condenses, forms clouds, and moves over land. Meteorologists use the term precipitation to describe any form of water which falls from the sky, whether it is snow, rain, or hail. If the precipitation falls to earth as rain it could run into a river, be absorbed into the ground, or plants might use it to help them grow. Water which runs into a river eventually flows to the ocean to begin the cycle again. If the water is absorbed into the ground in a process called infiltration, it can be used for agriculture, human consumption, or flow underground back into the ocean.

If the precipitation is frozen, it can fall to the ground as snow or become part of a glacier. The snow will melt when the temperature rises in the spring and the water will follow the same path as rain flowing into rivers. If the water becomes part of a glacier it may stay frozen for thousands of years. Recently, scientists have become concerned about melting glaciers around the world returning large amounts of freshwater to the ocean and upsetting the balance of the hydrologic cycle.

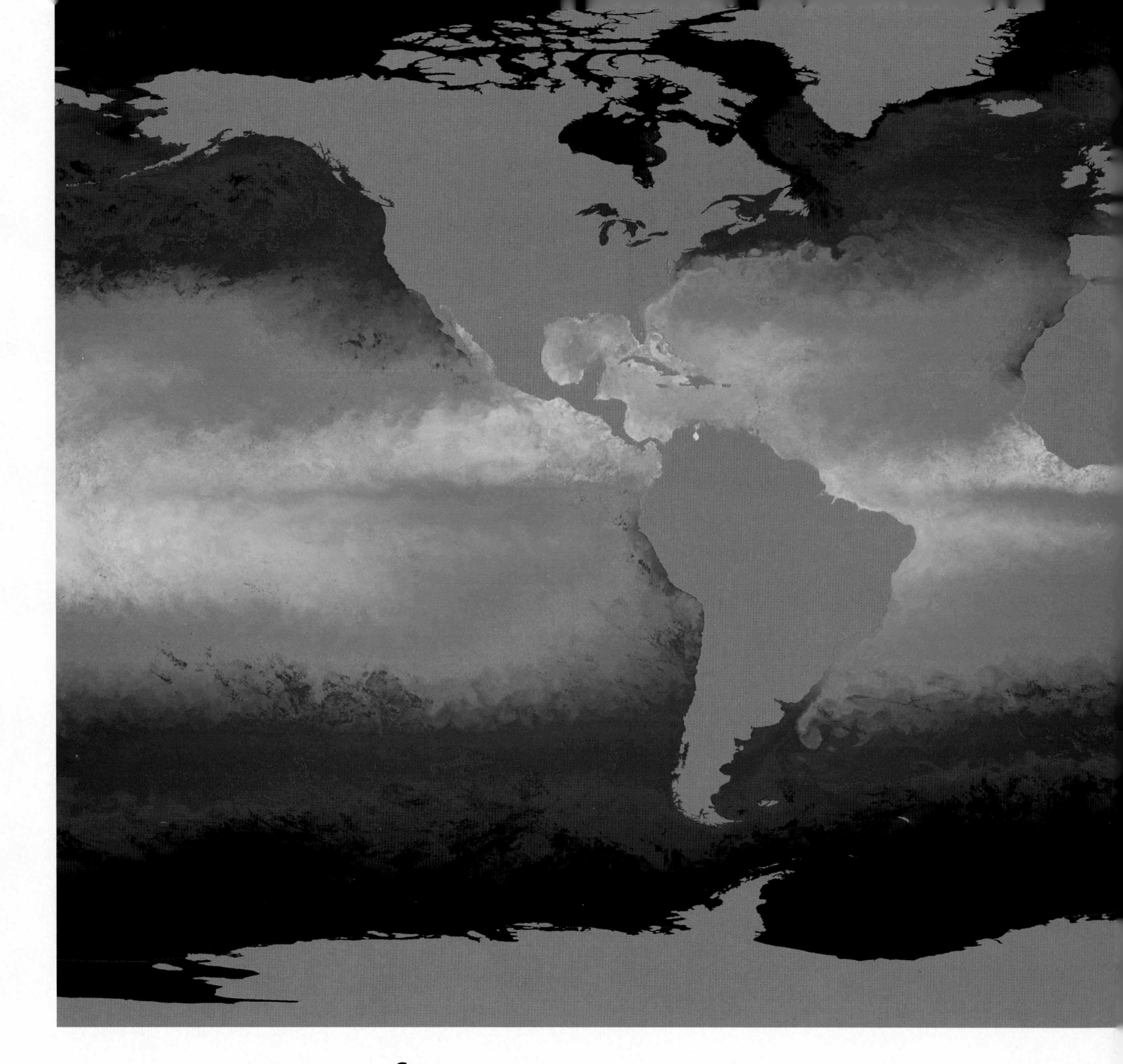

Sea surface temperature

The temperature of the ocean surface influences the lives of people all over the world. Knowledge of the sea surface temperature (SST) allows fishermen to target certain species of fish, captains to find currents to speed up their ships, and meteorologists to predict the path of major storms like hurricanes. Since the ocean stores large amounts of heat, repeated SST measurements allow scientists to observe long-term climate changes.

In the past, people disagreed on SST because of the techniques used to measure it. The original method of taking SST was by using a mercury thermometer and measuring the temperature of water obtained by lowering a bucket over the side of a ship. Advances in technology replaced this method with the automatic measurement of water by an instrument on the hull of a ship. Scientists considered both of these methods flawed because the depths of the measurements were not always consistent from vessel to vessel and the same location was not sampled regularly. Moored buoys give repeated measurements at the same depth, but they only sample one location in the entire ocean. A worldwide network of buoys monitors the ocean, yet not even all of them can give the spatial coverage necessary for the needs of many people.

In the early 1980s, scientists celebrated when reliable satellites began to measure global SST. These satellites measure the infrared radiation emitted by the ocean

surface. They are widely applauded because of their ability to measure large areas of the ocean at one time and the regularity at which they can take the measurements. Unfortunately, these valuable instruments have several weaknesses. The satellites can only measure the top one thirtieth inch (1 mm) of the ocean surface, and temperature may be very different a short distance below the surface due to cooling by the wind or heating by the daily passage of the sun. Also, the satellites cannot see infrared radiation through clouds. The next generation of satellites uses cloud penetrating microwave radiation to overcome this limitation.

ABOVE:
Colored computer model of global sea temperatures based on satellite data taken in June 2001. It is winter in the northern hemisphere and summer in the southern hemisphere. The surface temperature of the earth's oceans has been color-coded and combined with a projection of the land surface (grey). The temperature varies from a warm 95°F (35°C) (yellow) in the tropics, through red, blue, purple, and green to a freezing 28°F (-2°C) (black) in the polar regions.

Water movement and circulation

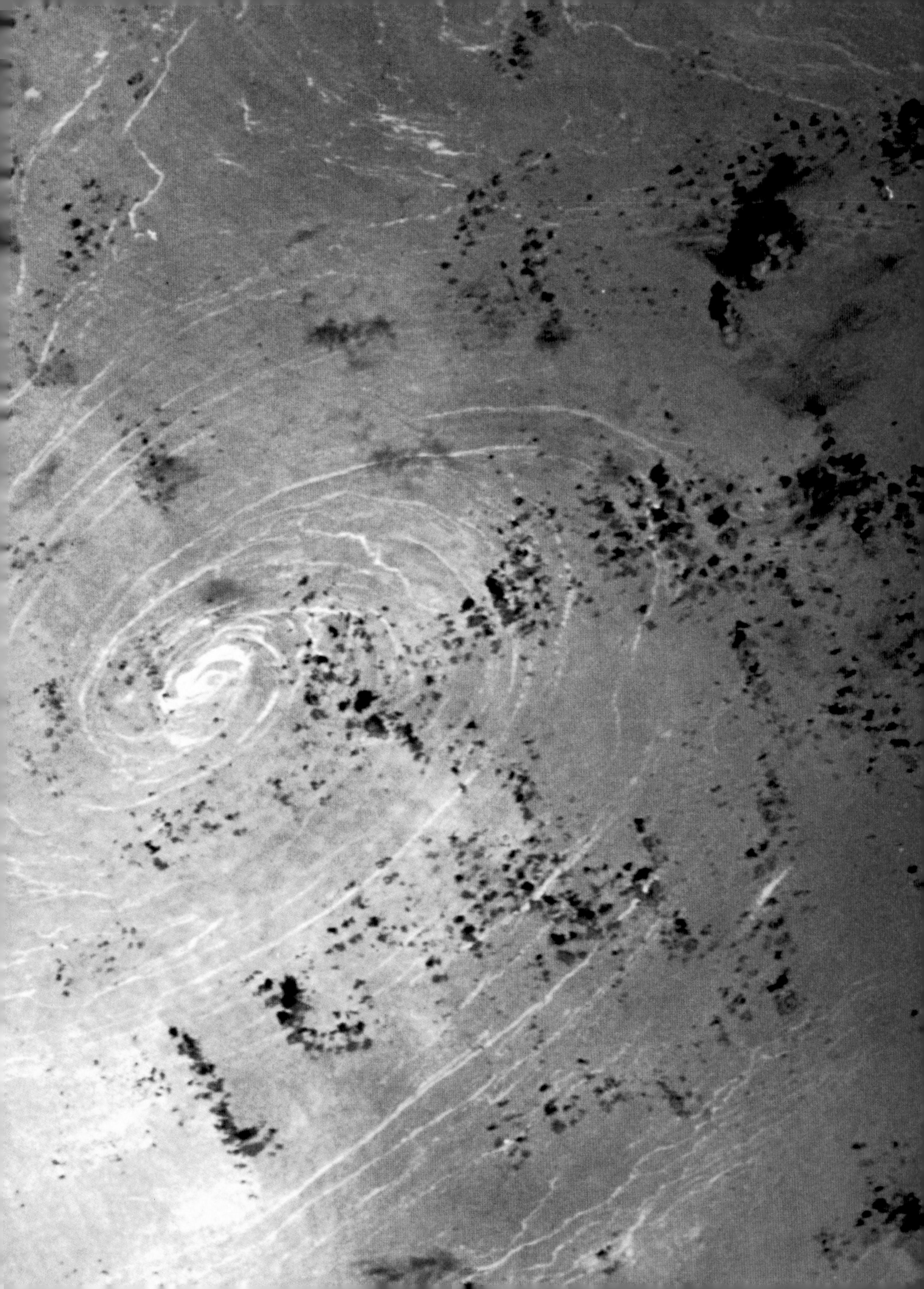

Water movement and circulation
Introduction

Winter assaults Nikolskoye, a Russian village on Bering Island in the north Pacific Ocean. During February, air temperature averages 27.3°F (–2.6°C), and one out of 100 days of the year is colder than 17.1°F (–8.3°C). At Port Ellen, a Scottish town on the Isle of Islay in the north Atlantic Ocean, the average February air temperature is 39.7°F (4.3°C) and only one out of 100 days averages below freezing (32°F, 0°C). But Port Ellen and Nikolskoye are both between 55°N and 56°N, so both are nearly the same distance from the north pole, and both get the same amount of sunlight on a cloudless winter day. Why is Nikolskoye so much colder?

The answer lies in the complex interaction of the sun, the moon, and the earth which drives oceanic and atmospheric circulation. At the earth's surface, we recognize the atmospheric circulation as the winds – ranging in scale from brief gusts to the great, ever-present wind systems blanketing various latitude bands. Ocean circulation consists of the various regional currents, carrying vast quantities of water and heat.

The Gulf Stream, a major current of the Atlantic Ocean, runs up the east coast of the United States from Florida to North Carolina. It carries thirty million tons per second of warm tropical water northward past Miami, Florida. After leaving the coast at Cape Hatteras, the Gulf Stream flows eastward towards Europe. Cold, dry air from Canada blows eastward over this warm water, pushing the current across the ocean and absorbing its tropical heat before reaching Scotland. In contrast, at Bering Island, the winds have passed over the bitter Siberian plains, and the coastal current flows from the north, carrying frigid Arctic water past the huddled village of Nikolskoye.

Two types of force are ultimately responsible for driving ocean circulation: heating by the sun, and the gravitational pull of the moon and sun. Solar heating creates the large-scale wind patterns which drive the oceans' surface currents, and warms the tropical waters near the equator. Cooling and freezing at high latitudes (near the north and south poles) causes water to sink beneath the surface, where it flows back toward the tropics. This water must eventually return to the surface. For this, the cold water must mix with warmer water above, becoming steadily lighter and shallower over a thousand years. Tidal motion, the back-and-forth sloshing pulled by the moon and sun's gravity, helps drive this mixing.

PREVIOUS PAGE:
Space Shuttle picture of open ocean turbulence in the Caribbean Sea between the Panama Canal and Jamaica. The interaction between two adjacent spiral eddies has produced a number of smaller spin-off eddy currents.

RIGHT:
The Gulf Stream is one of the Earth's strongest ocean currents which takes heat from the tropics far into the north Atlantic. Here it pulls away from the coast of the U.S. where the current widens, heads northeastward, and begins to meander. In the sea surface temperature image (bottom), the warm waters snake from bottom left to top right, showing several deep bends in the path, while the cold waters dip southward. Often chlorophyll, (top) which indicates the presence of marine plant life, is higher along boundaries between cool and warm waters, where currents are mixing up nutrient-rich water from deep in the ocean.

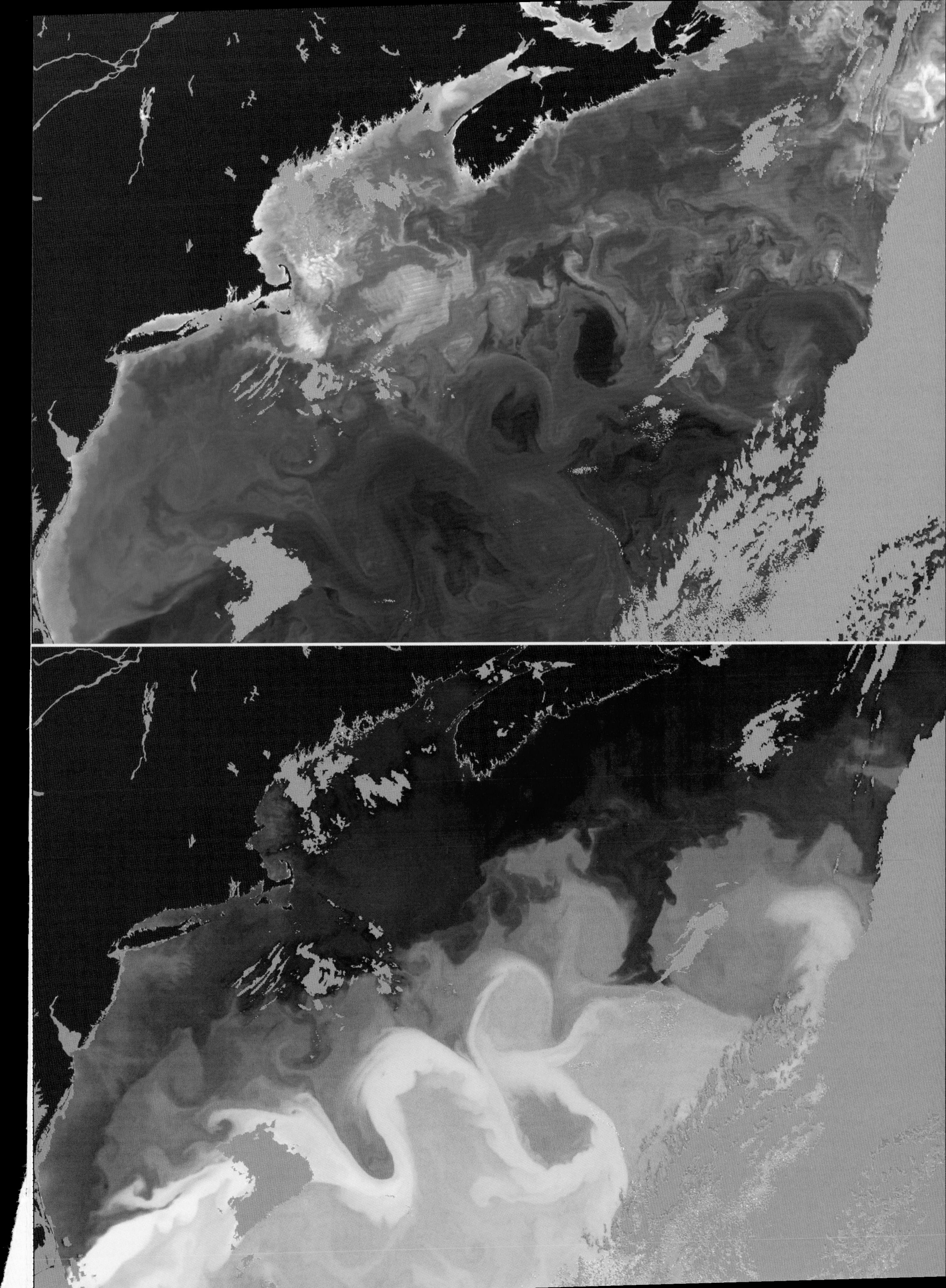

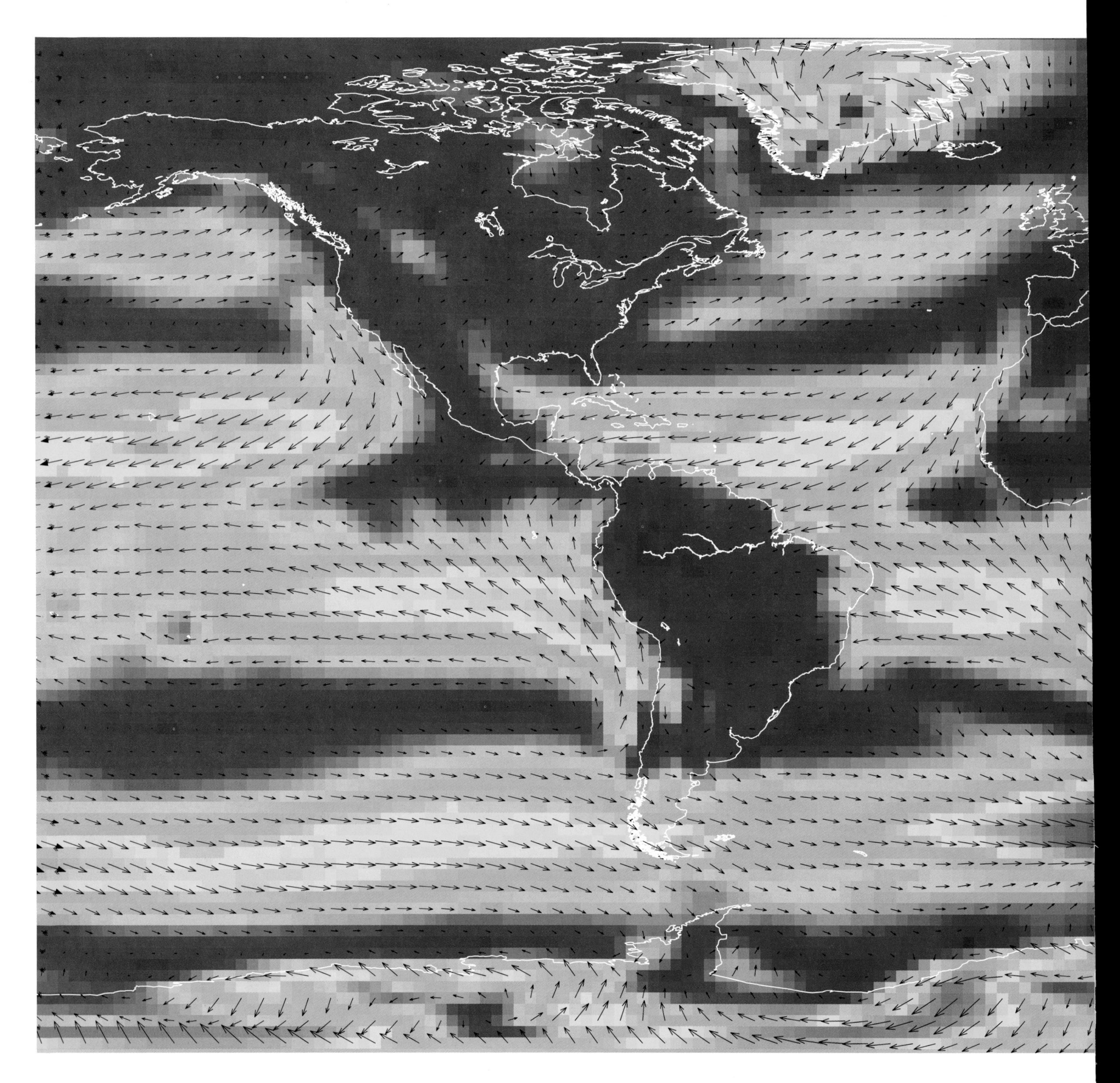

Surface winds

The sun shines directly above the tropics, warming water and air, while it shines obliquely across the poles. This heating difference causes air to rise near the equator, in a latitude band called the "Doldrums," and sink near the poles.

As the air rises in the Doldrums, it is replaced by air flowing from the north and south. Because the earth is rotating beneath this flow, the path taken by the air bends until it comes from the northeast (to the north of the Doldrums) or the southeast (to the south). This easterly (from the east) flow of air is called the Trade Winds.

After rising from the earth's surface at the Doldrums, the warm tropical air moves toward the poles and becomes steadily cooler and heavier. At around 30°N and S, the air sinks back to the surface. Surface winds in this latitude band, known as the Horse Latitudes, are generally weak and variable.

At the north and south poles, cold dense air moves toward the equator. Like the Trade Winds, the path of this air is bent by the earth's rotation until it comes from the east. At around 50°N and S, this air becomes warm enough to rise again.

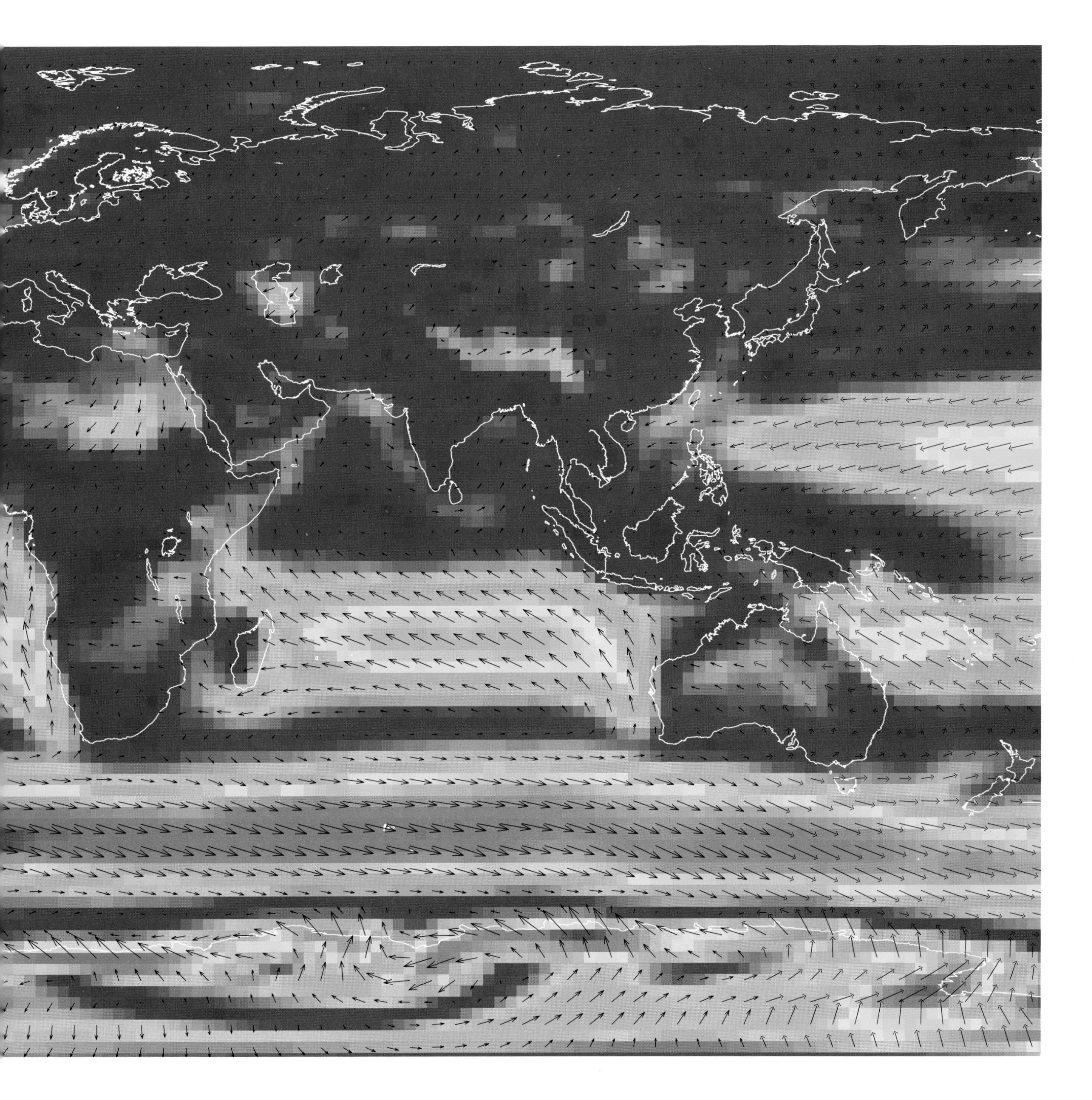

Between 30° and 50°N and S, air moves across the earth's surface from the sinking (30°) to the rising (50°), poleward. The earth's rotation bends this wind until it comes from the west. This band of wind is called the Westerlies.

feet per second
0.0 3.9 7.9 11.8 15.7 19.7 23.6 27.6 31.5 35.4 39.4
0.0 1.2 2.4 3.6 4.8 6.0 7.2 8.4 9.6 10.8 12.0
meters per second

ABOVE:
Average surface winds during 2006, showing the speed of the winds, and the direction (black arrows).

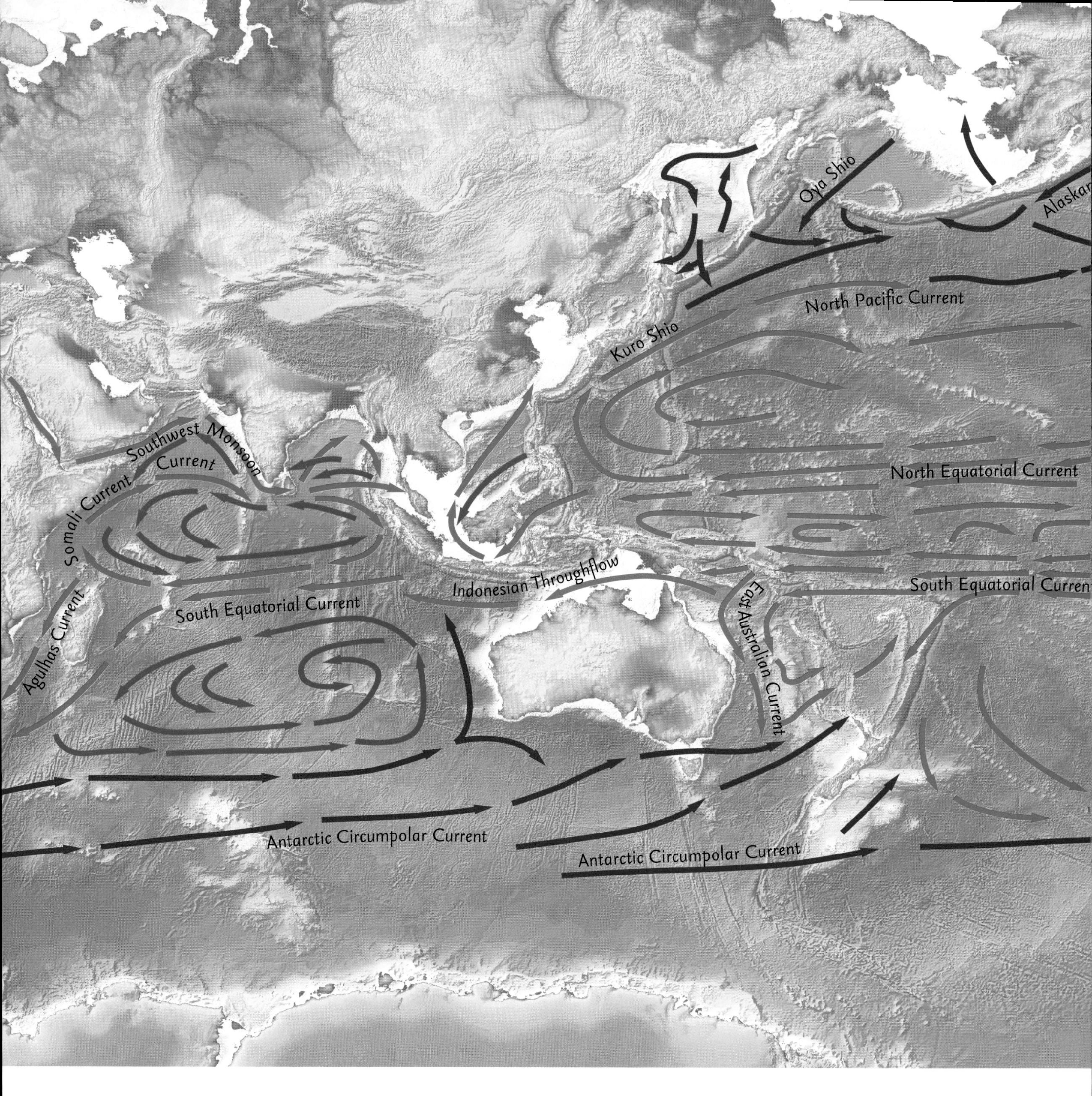

Surface currents

The ocean surface is pushed eastward by the Westerlies and westward by the Trade Winds. This large-scale "twist" causes water to pile up in the middle and swirl in vast, ocean-spanning gyres, clockwise in the northern hemisphere and counterclockwise in the southern hemisphere. Because the earth's surface is curved and rotating, water in the middle of these gyres tends to move toward the equator. This makes the gyres weaker on their eastern sides than on their western sides, where the water returns poleward in the strong western boundary currents: the Gulf Stream, Kuro Shio, Agulhas, Brazil, and East Australian Currents.

South of South America, a continuous, circumpolar band of ocean separates Antarctica from the other continents and connects the Pacific, Indian, and Atlantic Oceans at their southern boundaries. Here, the fierce southern Westerly winds drive the earth's most powerful current : the Antarctic Circumpolar Current, which carries 120–140 million tons of water each second through the Drake Passage separating South America from Antarctica.

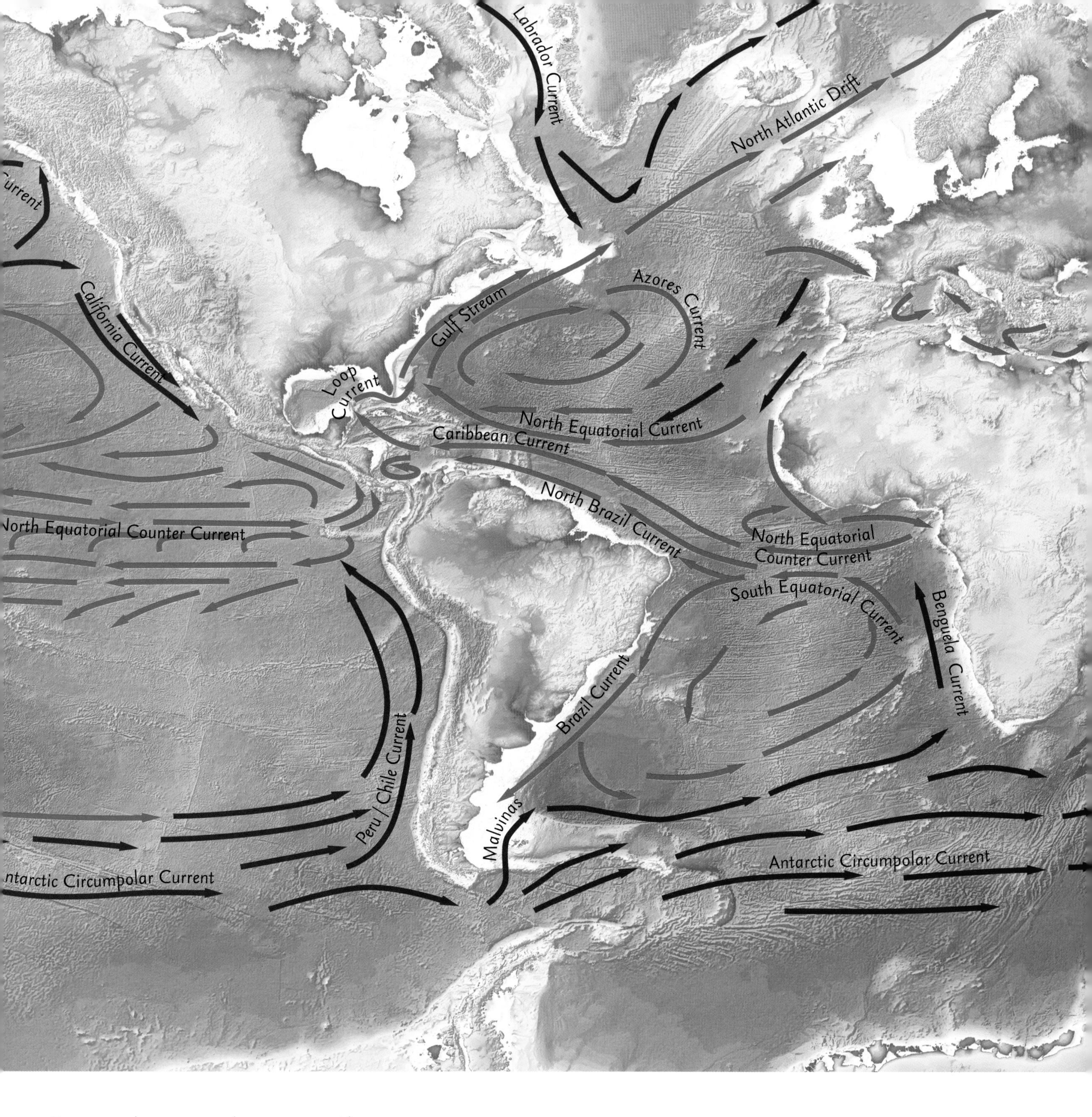

Oceanographers measure these currents with instruments moored (anchored) at fixed places in the ocean, and with buoys which float along with the currents. Many of these instruments send their data to passing satellites each day; and this stream of data is used for weather prediction, climate monitoring, and oceanographic research.

ABOVE:
The world's major surface currents, mapped using satellite-tracked surface drifting buoys.

Cold ocean current
Warm ocean current
Seasonal drift during northern winter

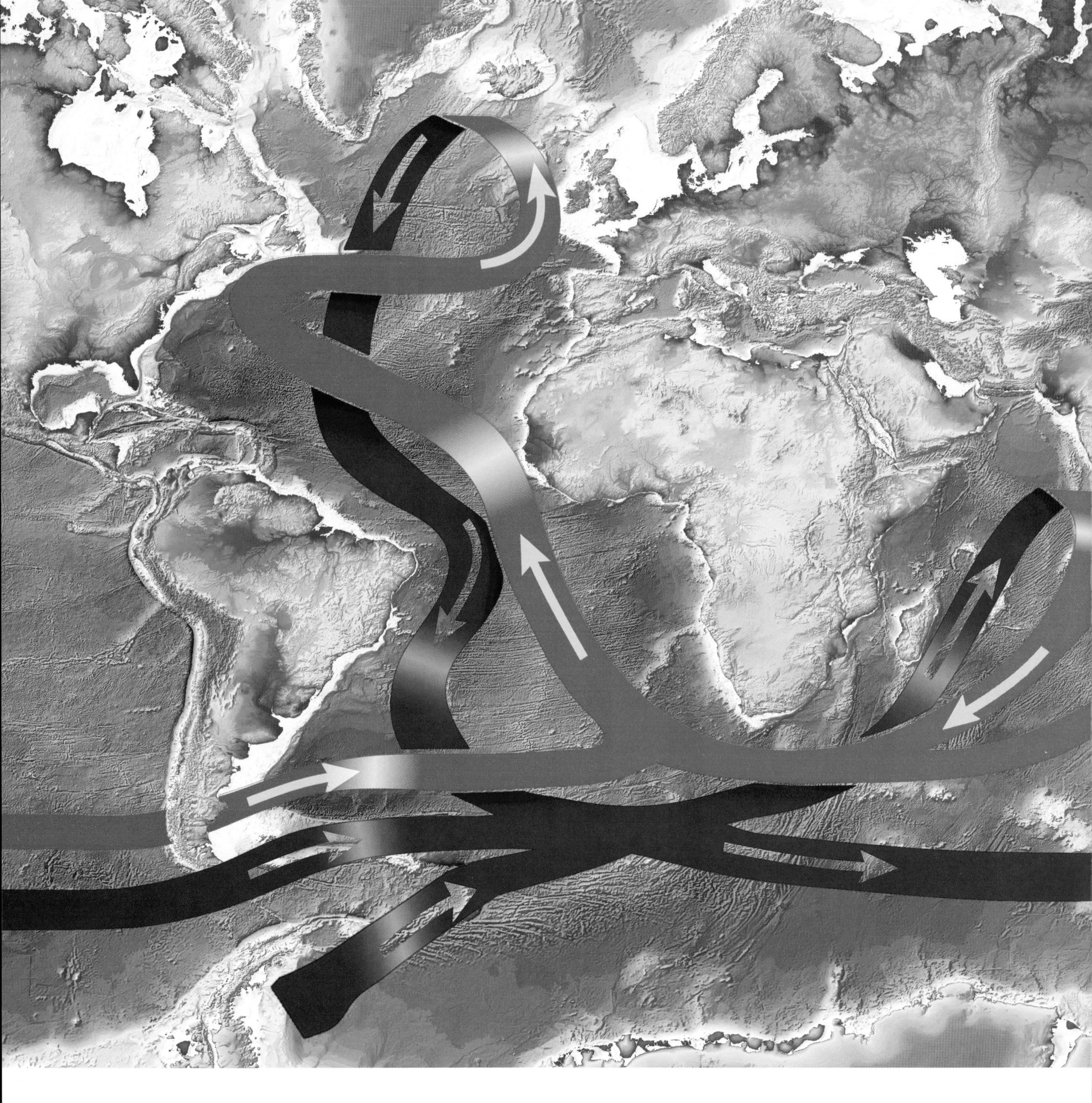

Ocean conveyor belt

Ocean currents are not limited to the ocean's surface. After the warm, salty tropical water of the Gulf Stream is carried to the north Atlantic, it cools, becomes extremely dense, and sinks beneath the surface. This sinking takes place in a few key places: the Nordic Seas north of Iceland, and the Irminger, and Labrador Seas surrounding southern Greenland. The deep water then returns southward in a deep western boundary current beneath the Gulf Stream. Sinking also happens in marginal seas against Antarctica, where the air is intensely cold and the water becomes extremely salty during ice formation. This Antarctic Bottom Water fills the deep Southern Ocean and spreads northward.

All of this dense bottom water must eventually be changed back to light surface water – if not, the light surface water would disappear! There are two ways in which this can happen: the dense water can outcrop (reach the ocean surface) in the Southern Ocean, where warmer air and rain make it lighter as the wind blows it northward, or it can mix with lighter water above it in the ocean abyss. Recent observations indicate that this mixing is concentrated in regions of rough bottom topography, such as undersea valleys and mountain ridges.

Oceanographers estimate that a parcel of water can take around a thousand years to completely travel around the world in this global conveyor belt. As the water sinks in this conveyor belt, it absorbs some of the carbon dioxide which humans are releasing to the atmosphere – without the conveyor belt, global warming would be stronger.

Disruption of the conveyor belt circulation, for example due to global warming, could cause serious changes in regional climate.

ABOVE:
The ocean conveyor belt circulation. Warm, surface water (red) is moved around in the wind-driven circulation patterns. At high latitudes, this water becomes cold and dense; it sinks and becomes bottom water (blue). Mixing with water above slowly changes the bottom water back to surface water.

Tides

The waters in the ocean basins are set into daily rhythmic motion in response to the gravitational interaction of the earth, moon, and sun. The basic movements are the earth spinning (rotating) daily on its axis, the moon revolving once per month around the earth, and the earth revolving once per year around the sun. As a result, tides and tidal currents change amplitude on daily, monthly and yearly time scales. For instance, higher ranges of tide occur when the sun and moon are lined up together during new and full moons (spring tides) and lower ranges of tide occur when the sun and moon are at right angles to each other with respect to the earth (neap tides). Lunar (due to the moon) tides also vary monthly as the moon is closest (perigee) and furthest (apogee) away in its orbit. Solar (due to the sun) tides vary yearly as the earth is closest (perihelion) and furthest (aphelion) from the sun. Some of the highest (and lowest) tides occur when perigee occurs close to the same time as a spring tide. There are other changes in the tides due to the monthly and yearly changes in the declination (angle north or south of the equator) of the moon and sun.

Not all ocean basins respond similarly to the tide-producing forces; they vary depending on their depth and size. Tides in the open ocean are generally small–less than 3.3 ft (1 m) range of tide. However, as the tide progresses from the oceans onto the continental shelves and into the bays and estuaries, the ranges of tide and the strengths of the tidal currents can dramatically increase. Some estuaries enhance the incoming ocean tide because of their unique shape and depth, such as the Bay of Fundy on the Atlantic coast of Canada, with over 36 ft (11 m) range of tide, and Cook Inlet, Alaska, with over 30 ft (9 m) range of tide. In contrast, some shallow bays with restricted inlets, such as Laguna Madre, Texas and Pamlico Sound, North Carolina, have virtually no tide. Most coastal areas have two high tides and two low tides each day (semidiurnal tides). However many areas also have a unique response to tidal forcing, in that they generally only have one tide per day (diurnal tides in the northern Gulf of Mexico).

RIGHT:
Low tide (left) compared to high tide (right) in a quiet bay on Campobello Island, New Brunswick, Canada.

Mean sea level

The concept of mean sea level is illusive. This is because the sea surface is very dynamic, and always in motion over very short time scales due to waves, currents, the effects of wind, and atmospheric pressure changes. Mean sea level moves with the ocean tides and ocean circulation patterns, changing due to the effects of El Niño/La Niña patterns over the decades, and responding to climate change over the long term. Global mean sea level change is the overall change in volume of the oceans which causes the oceans to rise and fall over the geologic millennia due to the coming and going of the ice ages. The absolute global sea level rise over the long term is thought to be due to a combination of factors: thermal expansion of the ocean waters due to global warming – potentially 3.3–13 ft (1–4 m); melting of the mountain glaciers – potentially 1.6 ft (0.5 m); melting of the Greenland ice sheet – potentially 30 ft (7 m); and melting of the Antarctic ice sheets – potentially 230 ft (70 m). The Intergovernmental Panel on Climate Change (IPCC) consensus is that global sea level will rise about 16 in (40 cm) by 2100, not accounting for any significant ice sheet melting.

Relative sea level change is the change in sea level relative to the land along the coasts and is a combination of global sea level rise and the local vertical land movement. Tide station networks measure relative sea level change and show that relative sea level rise can be as much as 0.5–0.8 in (15–20 mm) per year in areas of regional land subsidence (e.g. coastal Louisiana) and also show that there can be areas of relative sea level fall at 0.5–0.8 in (15–20 mm) per year in the glacial fjords in southeast Alaska where the land is emerging from the sea after the weight of the mountain glaciers is removed as they melt away. After taking into account vertical land movement in a global set of long-term tide station records, research scientists have estimated that global sea level has risen at the rate of about 0.08 in (2 mm) per year over the last century. More recently, using continuous near-global coverage of the satellite altimeter missions, scientists have estimated the global sea level is rising at about 0.12 in (3 mm) per year using the last 20-year time period of altimeter data.

Wind and waves

RIGHT:
Waves crash over the rocky shoreline of Tsitsikamma National Park situated on South Africa's famous Garden Route.

FAR RIGHT:
Breaking wave on the west coast of Scotland.

The wind not only produces currents, it creates waves. As wind blows across the smooth water surface, the friction or drag between the air and the water tends to stretch the surface. As waves form, the surface becomes rougher and it is easier for the wind to grip the water surface and intensify the waves.

How big wind waves get depends on three things:

- Wind speed. The wind must be moving faster than the wave crests for energy to be transferred.
- Wind duration. Strong wind that does not blow for a long period will not generate large waves.
- Fetch. This is the uninterrupted distance over which the wind blows without significant change in direction.

After the wind blows for a while, the waves get higher from trough to crest, and both the wave length and period become longer. As the wind continues or strengthens, the water forms whitecaps and eventually waves start to break. This is referred to as a fully developed sea.

The waves in a fully developed sea outrun the storm that creates them, lengthening and reducing in height in the process. These are called swell waves, or simply referred to as "swell." Swells organize into groups, smooth and regular in appearance, which can travel thousands of miles unchanged in height and period.

The longer the wave, the faster it travels. As waves leave a storm area, they tend to sort themselves out with the long ones ahead of the short ones, and the energy is simultaneously spread out over an increasing area. As the waves close in on the coast, they begin to feel the bottom and their direction of travel might change due to the contour of the land. Eventually, the waves run ashore, increasing in height up to 1.5 times in deep water, finally breaking up as surf.

Oceans and the climate

Oceans and the climate
Introduction

Earth's changing climate is influenced by the constant interactions between the ocean and the atmosphere. The ocean is the memory of the climate system and is second only to the sun in effecting variability in the seasons and long-term climate change. It is estimated that the ocean has the potential to store 1,000 times more heat than the atmosphere and fifty times more carbon. 85 per cent of the rain and snow which water our planet comes directly from the ocean. Rising sea level is one of the most immediate impacts of climate change.

The ocean takes up heat in the summer, stores it, and releases it to the atmosphere in winter, often far from where it was absorbed. Wind blowing on the ocean surface drives a current system which transports heat in the upper layers. Thermohaline circulation (THC) drives heat transport in the deep ocean. THC is caused by differences in seawater density which arise from temperature (thermal) and salt (haline). Cooler water and saltier water are dense and sink; warmer and fresher water are less dense and rise. THC connects all the ocean basins together in one of the fundamental heat engines of our planet.

Because the ocean is constantly in motion, the patterns of heat uptake and release vary from year to year, causing the seasons to vary from year to year. El Niño is perhaps the best-known ocean phenomenon affecting the seasons. When El Niño occurs, the equatorial Pacific Ocean becomes unusually warm, and this warming changes the patterns of atmospheric temperature, precipitation, and storms around the world.

Understanding the global carbon cycle is of critical importance to international policy decisions as well as to forecasting long-term trends in climate. Projections of climate change are closely linked to assumptions about carbon dioxide (CO_2) feedback effects between the atmosphere, the land, and the ocean. The ocean stores vast amounts of carbon and there is constant exchange of CO_2 between ocean and atmosphere. The equatorial Pacific is a strong source of CO_2 as a result of ocean waters upwelling to the surface. During El Niño years the equatorial Pacific source is lessened because upwelling is suppressed. During La Niña years the upwelling is enhanced and more CO_2 is released to the atmosphere. In the high-latitudes, the CO_2 exchange is governed primarily by deep currents rising to the surface and out-gassing in winter, and by sea life uptake of CO_2 in the summer. In the mid-latitudes, the air sea exchange of CO_2 is governed primarily by water temperature; summer temperatures cause out-gassing and winter temperatures cause the ocean to take up CO_2.

Rising sea level is one of the most visible consequences of climate change. Sea level rise results from two sources. First, a warming atmosphere causes continental ice sheets and glaciers to melt, adding more water to the ocean. Second, as the ocean gets warmer, the water increases in volume due to thermal expansion, and the volume increase causes sea level to rise. During the twentieth century, sea level rose at a rate of about 0.04 inches (1.0 mm) per year in the first half of the century, and the rate increased to about 0.08 inches (2.0 mm) per year in the second half. Recent measurements indicate that over the past ten years, the rate of sea level rise has increased to about 0.12 inches (3.1 mm) per year. Scientists estimate that about half of the rise is due to melting ice and about half of the rise is due to thermal expansion.

RIGHT:
Storm clouds at sunset. Changes in the ocean influence changes in the patterns of storms and other extreme weather events.

RIGHT:
Warming of surface waters in the southeastern Pacific creates the phenomenon known as El Niño. It creates severe winter weather and large ocean swells.

PREVIOUS PAGE:
This spectacular "blue marble" image is the most detailed true-color image of the entire earth to date and shows the world's oceans and weather patterns in amazing detail.

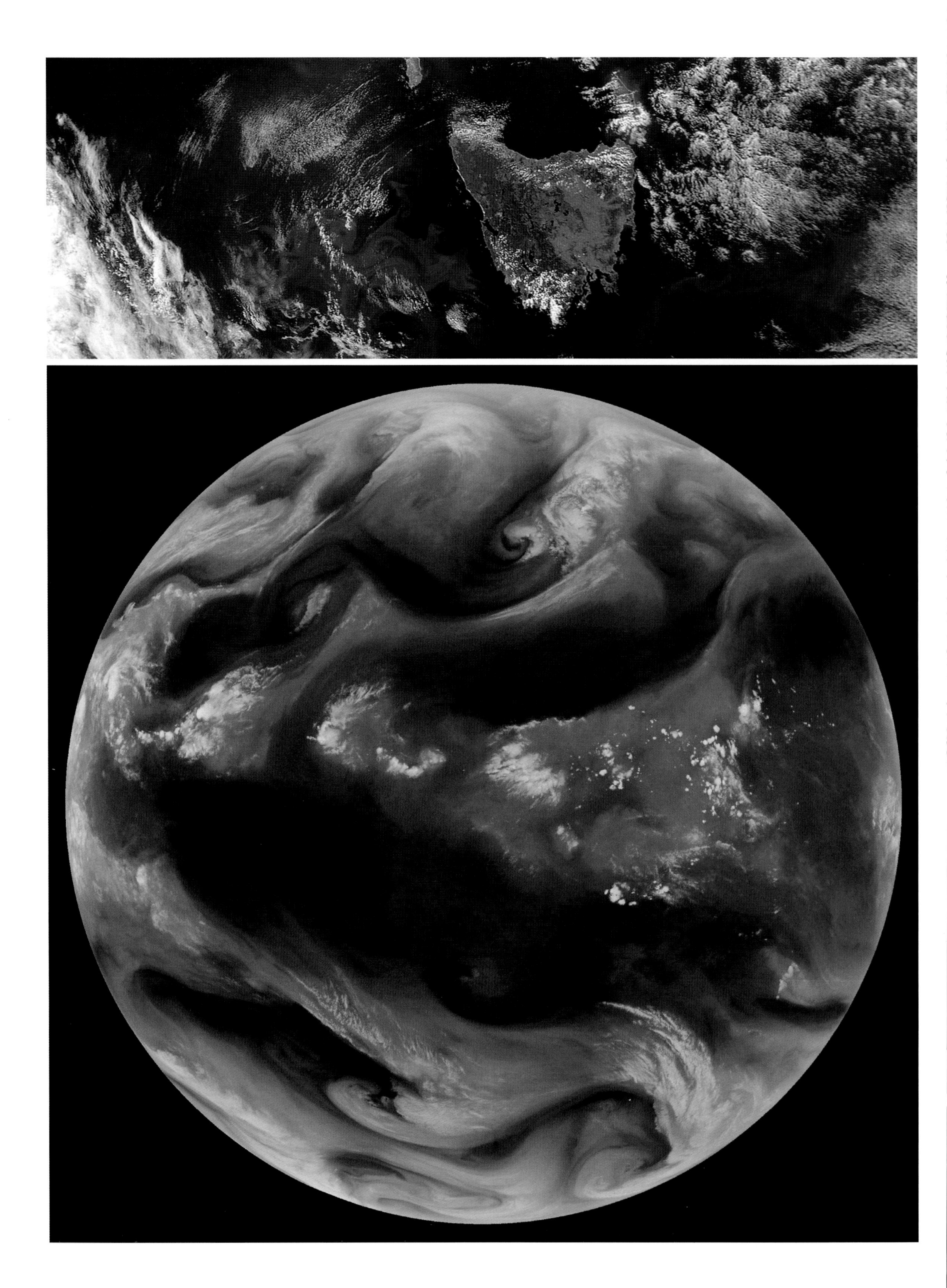

Ocean circulation climate modifier

The oceans are the great moderator of climate throughout the world. They respond largely to winds, the winds respond to the distribution of heat, which in turn depends on ocean circulation. The general circulation of the ocean and that of the atmosphere are closely connected in two main ways.

The first is physically, through the exchange of heat, water, and momentum. Oceans cover more than 70 per cent of the earth's surface, contain around 97 per cent of its surface water, and store enormous amounts of heat. The heat rises from the oceans' surfaces and warms the atmosphere. As this takes place, air temperature gradients are created and, as a consequence, winds. The winds push horizontally against the sea surface and drive ocean current patterns. At the same time, the variations in ocean temperature and salinity control the vertical ocean currents. Warm, fresh (less salty) waters flow upwards while colder, dense (saltier) waters tend to sink. Warm surface waters move poleward where the heat escapes more readily into outer space, while the cold, deep currents are established in the ocean depths. Through this ocean circulatory system, the oceans and atmosphere work together to distribute heat and regulate climate, so much so that some scientists have recently described the oceans as the "global heat engine."

The second way which the oceans interact with the atmosphere is chemically – the oceans are both a source and a sink of greenhouse gases. The heat which escapes from the oceans is mostly in the form of evaporated water, the most abundant greenhouse gas on earth. However, the water contributes to the formation of clouds, which shade the earth's surface and have a cooling effect. Scientists still do not know which process (cloud shading or water vapor heat retention) will exert the larger influence on global temperatures.

Carbon dioxide (CO_2) is one of the most important greenhouse gases because of its links with human activities. Most of the world's carbon is found in the oceans. The processes which result in exchanges between the surface ocean and the atmosphere, and between the upper ocean and the deep ocean, are critical. Carbonate chemistry regulates much of the transfer between these systems, but biological processes, such as photosynthesis (which turns CO_2 into organic material), also play an important role. The settling of this organic carbon into the deep ocean is known as the "biological pump." Some scientists speculate that if the ocean's circulation pattern is disrupted, it could become a source rather than a sink for carbon and atmosphere CO_2 levels could rise much higher than they are now.

Although scientists have developed sophisticated climate models to help them understand the oceans' role in moderating climate, many questions still remain unanswered. One of the key questions in climate change research is how the physical and biological processes of the oceans will respond to chemical and physical changes in the atmosphere.

TOP LEFT:
This image from NASA's satellite-borne Sea-viewing Wide Field-of-view Sensor (SeaWiFS) shows color differences caused by large plankton blooms in the ocean. The plankton blooms in turn delineate eddy patterns in the regional ocean circulation southeast and southwest of Tasmania.

LEFT:
Picture of the earth taken by Meteosat showing water vapor in the atmosphere. For over two decades Europe's Meteosat satellites have been sending pictures of the weather back to the earth for viewing by millions on daily TV weather bulletins.

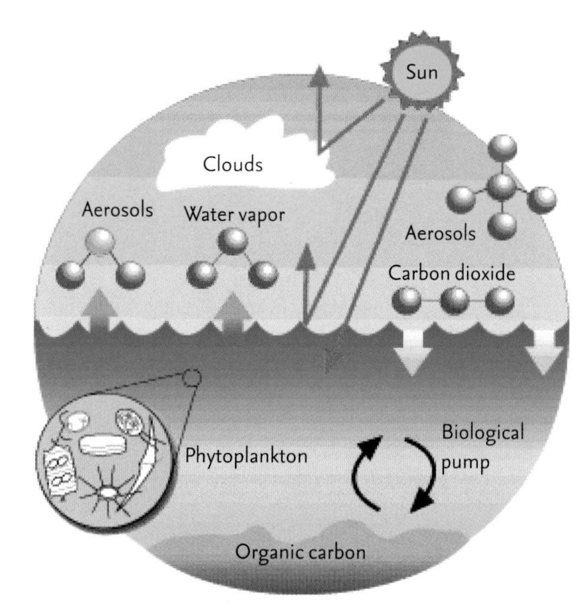

LEFT:
Atmospheric circulation.
The ocean has been dubbed the "global heat engine." Energy escapes the ocean in the forms of heat and water vapor. As the atmosphere warms, temperature gradients are created, resulting in surface winds which, in turn, drive ocean currents. These winds and water vapor also dramatically affect meteorological conditions, resulting in the formation of clouds or even rainstorms which are vital for life on land.

Inter-Tropical Convergence Zone (ITCZ)

The Inter-Tropical Convergence Zone (ITCZ), appears as a band of clouds, usually thunderstorms, which circle the globe near the equator. The solid band of clouds may extend for many hundreds of miles and is sometimes broken into smaller line segments. The ITCZ follows the sun in that its position varies seasonally. It moves north in the northern summer and south in the northern winter. The ITCZ is what is responsible for the wet and dry seasons in the tropics.

It exists because of the convergence of the trade winds. In the northern hemisphere the trade winds move in a southwesterly direction, while in the southern hemisphere they move northwesterly. The point the trade winds converge forces the air up into the atmosphere, forming the ITCZ.

Thunderstorms in the tropics tend to be short, but can produce intense rainfall. Greatest rainfall typically occurs when the midday sun is overhead. On the equator this occurs twice a year in March and September, and consequently there are two wet and two dry seasons.

Further from the equator, the two rainy seasons merge into one, and the climate becomes more monsoonal, with one wet season and one dry season. In the northern hemisphere, the wet season occurs from May to July, in the southern hemisphere from November to February.

Tale of Two Cities: Kano and Lagos

Nigeria's climate is characterized by the hot and wet conditions associated with the ITCZ north and south of the equator. This is easily seen in the normal monthly rainfall for two cities, Kano and Lagos, separated by 500 miles (800 km).

When the ITCZ is to the south of the equator, the northeast winds prevail over Nigeria, producing the dry-season conditions. When the ITCZ moves into the northern hemisphere, the south westerly wind prevails as far inland to bring rainfall during the wet season. This results in a prolonged rainy season in the far south of Nigeria, while the far north undergoes long dry periods annually. Nigeria, therefore, has two major seasons, the dry season and the wet season, the lengths of which vary from north to south.

In southern Nigeria, Lagos averages 68.5 inches (1,740 mm) of rain annually. The four observed seasons are:

- The long rainy season from March to the end of July, with a peak period in June over most parts of southern Nigeria.
- The short dry season in August lasting for 3–4 weeks. This is due to the ITCZ moving to the north of the region.
- The short rainy season following the brief wet period in August and lasting from early September to mid-October as the ITCZ moves south again, with a peak period at the end of September. Rains are usually lighter than the long rainy season.
- The long dry season from late October to early March with peak dry conditions between early December and late February.

In northern Nigeria, Kano averages 32.5 inches (825 mm) of rain annually. There are two observed seasons since the ITCZ only moves into the region once a year before returning south. These are:

- The long dry season from October to mid-May. With the ITCZ in the southern hemisphere, the northeast winds and their associated easterlies over the Sahara prevail over the country, bringing dry conditions. There is little or no cloud cover.
- The short rainy season from June to September. Both the number of rain days and total annual rainfall decrease progressively from the south to the north. The rains are generally heavy and short in duration, and often characterized by frequent storms.

ABOVE:
This image is a combination of cloud data from NOAA's Geostationary Operational Environmental Satellite (GOES-11) and color land cover classification data. The ITCZ is the band of bright white clouds which cuts across the top half of the image.

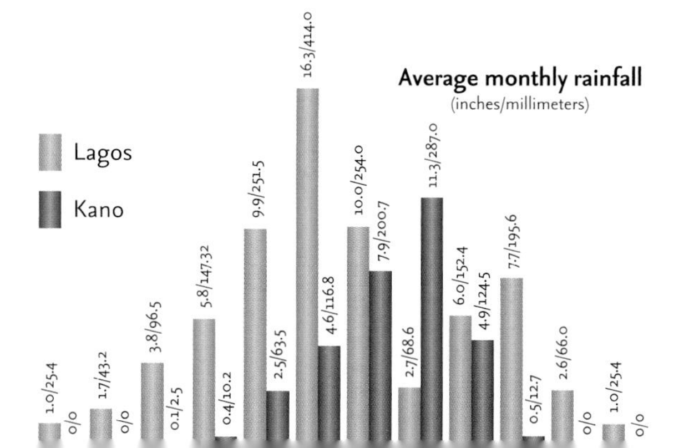

El Niño and La Niña

El Niño is an unusual warming of the tropical Pacific Ocean which occurs irregularly once every two to seven years. In the nineteenth century, Peruvian fishermen gave the name "El Niño"—with specific reference to the Christ Child—to a warm current which appeared off their coast around Christmas time. In the 1960s, scientists discovered that unusual warming off Peru during El Niño years typically extends 6,000 miles (9,660 km) further to the west across the Pacific basin.

El Niños develop because of a "chicken and egg" feedback loop between the ocean and the atmosphere. A weakening of the trade winds along the equator causes ocean temperatures to rise; then the elevated ocean temperatures cause the trade winds to weaken even further. These interactions lead to El Niño warming which typically lasts for about a year.

Changes in ocean temperatures and currents during El Niño reduce the growth of one-celled plants called phytoplankton at the base of the marine food chain. These effects on phytoplankton reverberate through the entire ecosystem in the Pacific. Fish, marine mammals, and seabirds may starve, fail to reproduce, or migrate to different parts of the ocean in search of food. For this reason, El Niño can lead to changes in the geographic distribution and catch of commercially valuable fish species such as tuna and Peruvian anchovy.

During El Niño, the ocean pumps a tremendous amount of heat and moisture into the tropical atmosphere. Global wind systems, including the high-altitude jet streams, respond to these oceanic forces, causing worldwide shifts in weather patterns. As a result, El Niño increases the likelihood of droughts, floods, heat waves, and extreme weather events in many regions of the globe both near and far from the tropical Pacific. El Niño-related rainfall and flooding can also contribute to the spread of insect- and water-borne diseases like malaria, dengue fever, cholera and dysentery. It is estimated that weather-related disasters during the 1997–98 El Niño claimed 22,000 lives and caused $36 billion in economic losses worldwide.

However, El Niño is not all bad news. For instance, El Niño often suppresses the formation of deadly Atlantic hurricanes, which reduces loss of life and damage to coastal communities. Also, El Niño winters are often milder in the midwest U.S., which benefits consumers by lowering home heating bills.

Many institutions around the world now routinely issue forecasts of El Niño warming in the Pacific and its impacts on global weather patterns. These forecasts are based on computer model predictions plus information on the behavior of previous El Niños. Forecasts are valuable for anticipating potential disasters or for taking advantage of El Niño's benefits.

In the late 1980s, scientists became aware of a cold counterpart to El Niño which is referred to as La Niña. During La Niña, trade winds strengthen and the tropical Pacific becomes unusually cold. Like El Niño, La Niña affects patterns of weather variability around the world in its own particular way. Sometimes El Niño and La Niña are referred to together as the El Niño/Southern Oscillation (ENSO) cycle of Pacific warm and cold events.

RIGHT:
Record high 35 foot (10 meter) surf due to El Niño on north shore of island of Oahu in the state of Hawaii U.S.A.

RIGHT:
Schematic of normal, El Niño, and La Niña conditions. The thermocline separates the warm sunlit surface layer from the cold ocean interior. Arrows represent the direction of winds and ocean currents. Coldest ocean surface temperatures are deep blue and the warmest ocean surface temperatures are red.

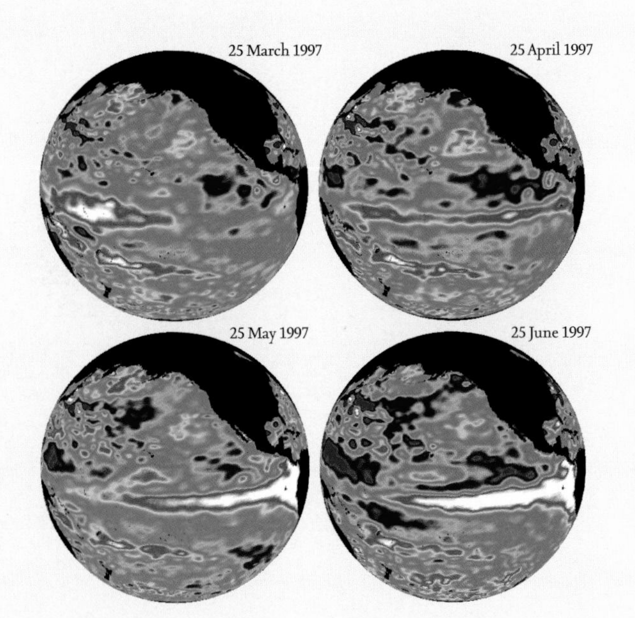

LEFT:
Weakening trade winds during El Niño cause pulses of high sea level which travel from west to east along the equator in the Pacific as shown in this sequence of satellite images between March and June 1997. On reaching the west coast of the Americas, these pulses (colored in red and white), in combination with storm generated surface waves and high tides, can contribute to coastal flooding and erosion during El Niño events.

FAR RIGHT:
The west coast of the U.S.A. is vulnerable to storm damage. These photographs were taken as part of a project to record coastal erosion resulting from severe storms generated as part of the Pacific Ocean warming effect of El Niño. In just one winter a significant amount of coastline has vanished leaving more buildings exposed.

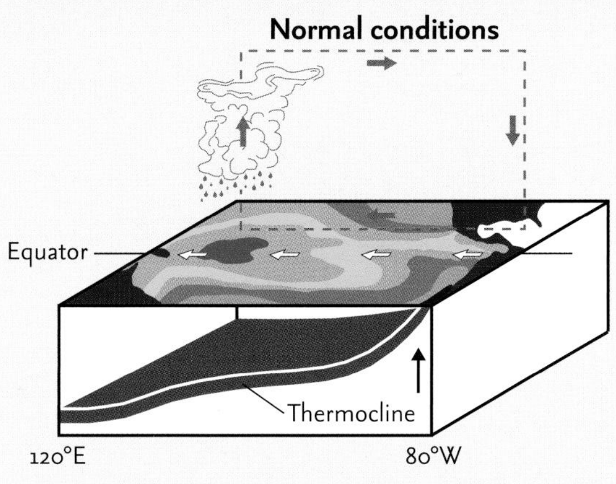

Normal conditions
Equator
Thermocline
120°E
80°W

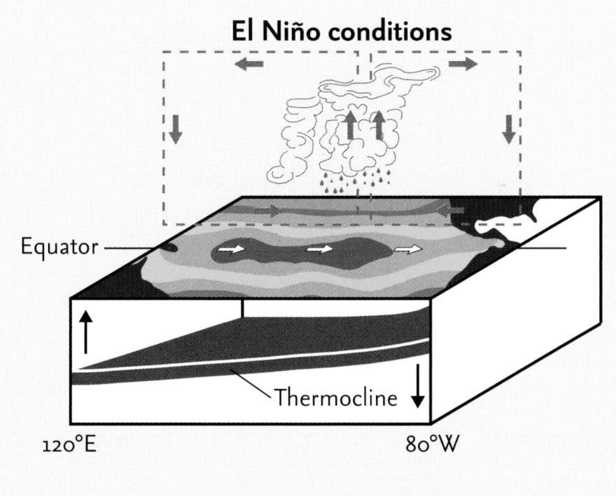

El Niño conditions
Equator
Thermocline
120°E
80°W

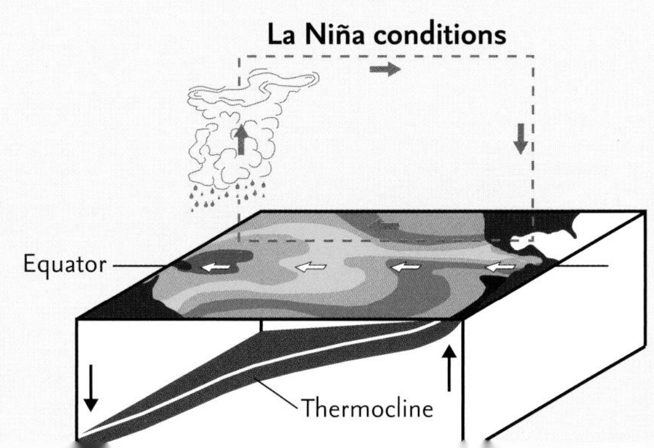

La Niña conditions
Equator
Thermocline

Oceans and the weather

Oceans and the weather
Introduction

The world's oceans comprise approximately 70 per cent of the total area of the globe. Because of this sheer dominance in terms of global coverage, they exert a major influence on weather and climate over the globe through the exchanges of energy and moisture in the overall earth/energy balance. Thus the world's oceans are "in effect," a giant factory for the production of tropical and extra-tropical storms, clouds, fog, land, and sea breezes, and related phenomena. Tropical cyclones, also called hurricanes, typhoons or cyclones depending on the geographical region, form over the tropics, gaining their energy from the upward fluxes of heat and moisture from the ocean surface. Extra-tropical cyclones of the middle and high latitudes, in general, also either develop over the ocean or intensify there due to reduced frictional drag and the supply of heat and moisture from the water surface.

Air masses are modified substantially as they pass over ocean waters. Warm moist air masses of tropical origin are cooled from below as they move pole-ward over colder water, producing sea fog and low clouds. Conversely, cold polar or arctic air masses are heated and supplied with moisture as they push towards the equator over progressively warmer water. This process generates upward movement of parcels of air, or convection, which produces cloud and showers.

Ocean currents are permanent or continuous directed movements of ocean water which flow in any of the earth's oceans. Some currents flow for thousands of miles and are generated from the forces acting upon the water like the earth's rotation, the wind, temperature and salinity differences, and the gravitational attraction of the moon. Ocean depth contours, the shoreline, and other currents influence a current's direction and strength. Ocean currents play a major role in weather and climate across the globe and are especially important in determining the climates of the continents, particularly those regions bordering the ocean. Perhaps the most striking example is the Gulf Stream, which carries warm water well into the mid-latitudes of the Atlantic Ocean and makes northwest Europe much more temperate than any other region at the same latitude. Conversely, the Hawaiian Islands are somewhat cooler than the tropical latitudes in which they are located because of the cooler waters carried by the north Equatorial Current.

The formation of extra-tropical cyclones in the middle latitudes and tropical cyclones over the tropical oceans are one of the key mechanisms by which the oceans play a role in the overall energy balance of the globe. The next sections describe in detail extra-tropical cyclones and tropical cyclones including such information as areas of geographic occurrence, relative size and intensity, and means to categorize these phenomena.

PREVIOUS PAGE:
An Atlantic frontal system approaches the island of Islay in western Scotland, U.K..

RIGHT:
NASA's Terra satellite captured this true-color image of Typhoon Namtheun on July 30, 2004. The Japanese coastline is clearly visible as the storm approaches it.

Tropical cyclones

A tropical cyclone is a generic term used to describe a cyclone which originates over the tropical or subtropical oceans with a counterclockwise circulation in the northern hemisphere and clockwise in the southern hemisphere. These systems are fueled by latent heat released by condensation of water vapor in showers and thunderstorms ("convection") near the system center. Because this mechanism makes the center of a tropical cyclone warmer than the outside, they are often referred to as warm-core systems. Tropical cyclones are distinguished from middle latitude cyclonic disturbances, more commonly referred to as extra-tropical storms, by the heat mechanism which fuels them. Extra-tropical storms derive their energy from temperature differences in the atmosphere and are usually cold-core systems. Depending on their location and strength, there are various terms by which tropical cyclones are known, such as hurricanes, typhoons or cyclones.

Tropical cyclones can produce extremely strong winds, tornadoes, torrential rain, high waves, and storm surges. They form and are sustained over large bodies of warm water with temperatures greater than 26.5oC (80oF) and lose their strength over land. This is why coastal regions can suffer significant damage from a tropical cyclone while inland regions for the most part are relatively safe from receiving strong winds. Heavy rains, however, can produce significant flooding inland, and storm surges can produce extensive coastal flooding up to 25 miles (40 km) inland. Although their effects on human populations can be devastating, tropical cyclones can also have beneficial effects, such as relieving drought conditions. They carry heat away from the tropics, an important mechanism of the global atmospheric circulation which helps maintain equilibrium in the earth's troposphere.

Tropical cyclones are initiated by a variety of disturbances including easterly waves, monsoon troughs, and depressions. Once initiated, tropical cyclones evolve through several stages in their development. The first stage, a tropical depression, is attained when a closed circulation with organized convection is noted with one minute sustained wind speeds of less than 34 knots (63 km/hr). As the system develops further and sustained winds increase to between 34–63 knots (63–119 km/hr), it is classified as a tropical storm. Hurricane, cyclone or typhoon status is reached when sustained winds exceed 64 knots (119 km/hr). Further classifications of systems once they reach hurricane, typhoon or cyclone status are basin-dependent and are described in detail in the subsequent sections.

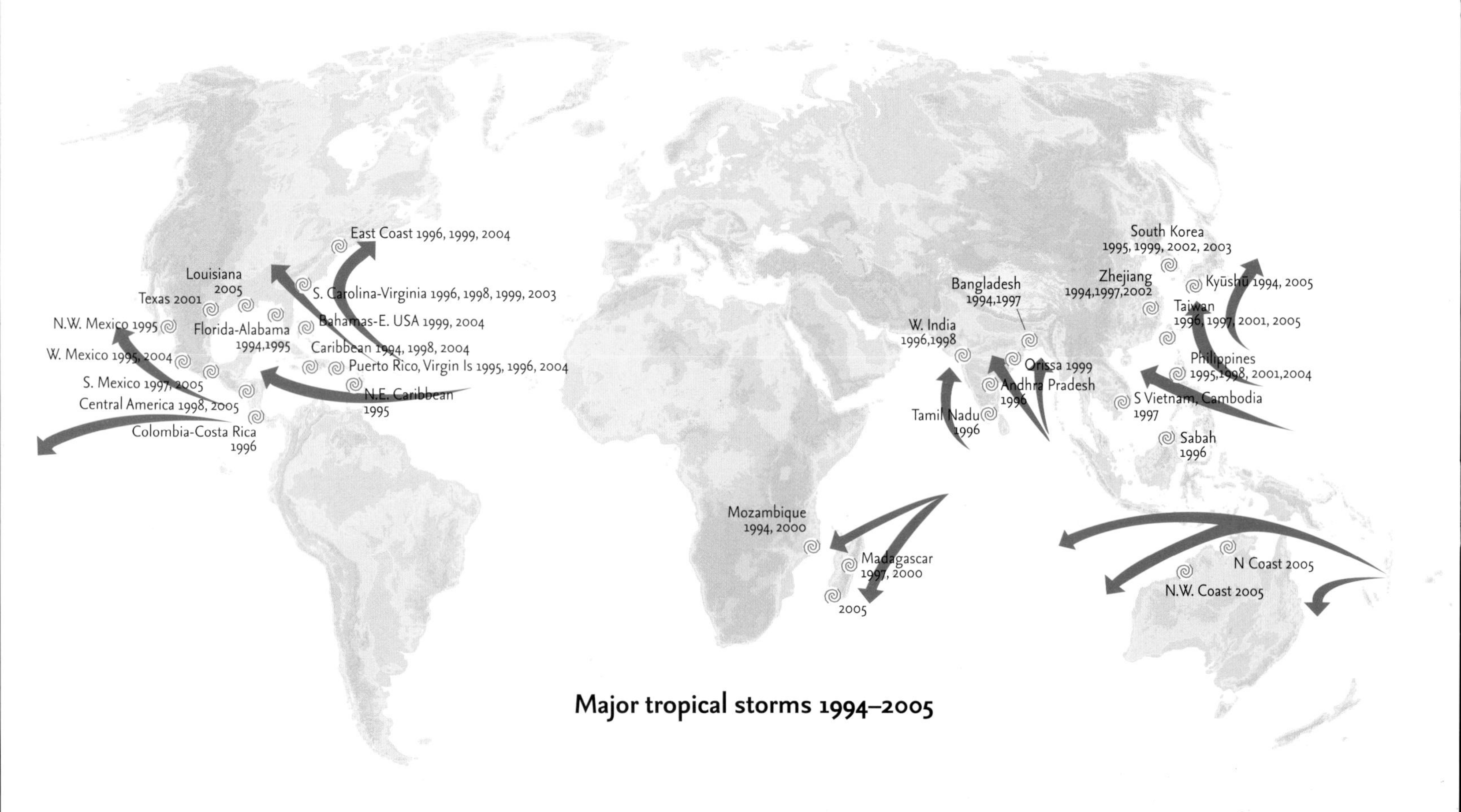

Major tropical storms 1994–2005

Subtropical cyclones

Another class of cyclone of note are the subtropical cyclones, which have characteristics of both tropical and mid-latitude cyclones. These systems initially derive their energy from temperature gradients in the atmosphere as an extra-tropical storm then gradually acquire tropical characteristics through convective redistribution of heat acquired from warm sea surface temperatures. Subtropical cyclones also differ from true tropical cyclones in that the thunderstorm activity or convection is not often concentrated near the center and therefore they typically have a larger radius of maximum winds.

Subtropical cyclones in the Atlantic basin are classified by their maximum winds in much the same manner as their tropical counterparts, with subtropical depressions possessing winds less than 34 knots (63 km/hr) and subtropical storms containing winds between 34–63 knots (63–119 km/hr).

Subtropical cyclones, or systems which develop through processes similar to north Atlantic subtropical cyclones, have been observed in the Great Lakes region of the U.S. (Hurricane Huron on September 14, 1996), the south Atlantic Ocean off the coast of South America (Cyclone Catarina in March, 2004, was probably a subtropical cyclone at one point during its development), the Mediterranean Sea, the north central Pacific Ocean (Kona lows near Hawaii), and the Indian Ocean near Madagascar (for example, Subtropical Cyclone Luma in April, 2003).

Cross section of a tropical storm

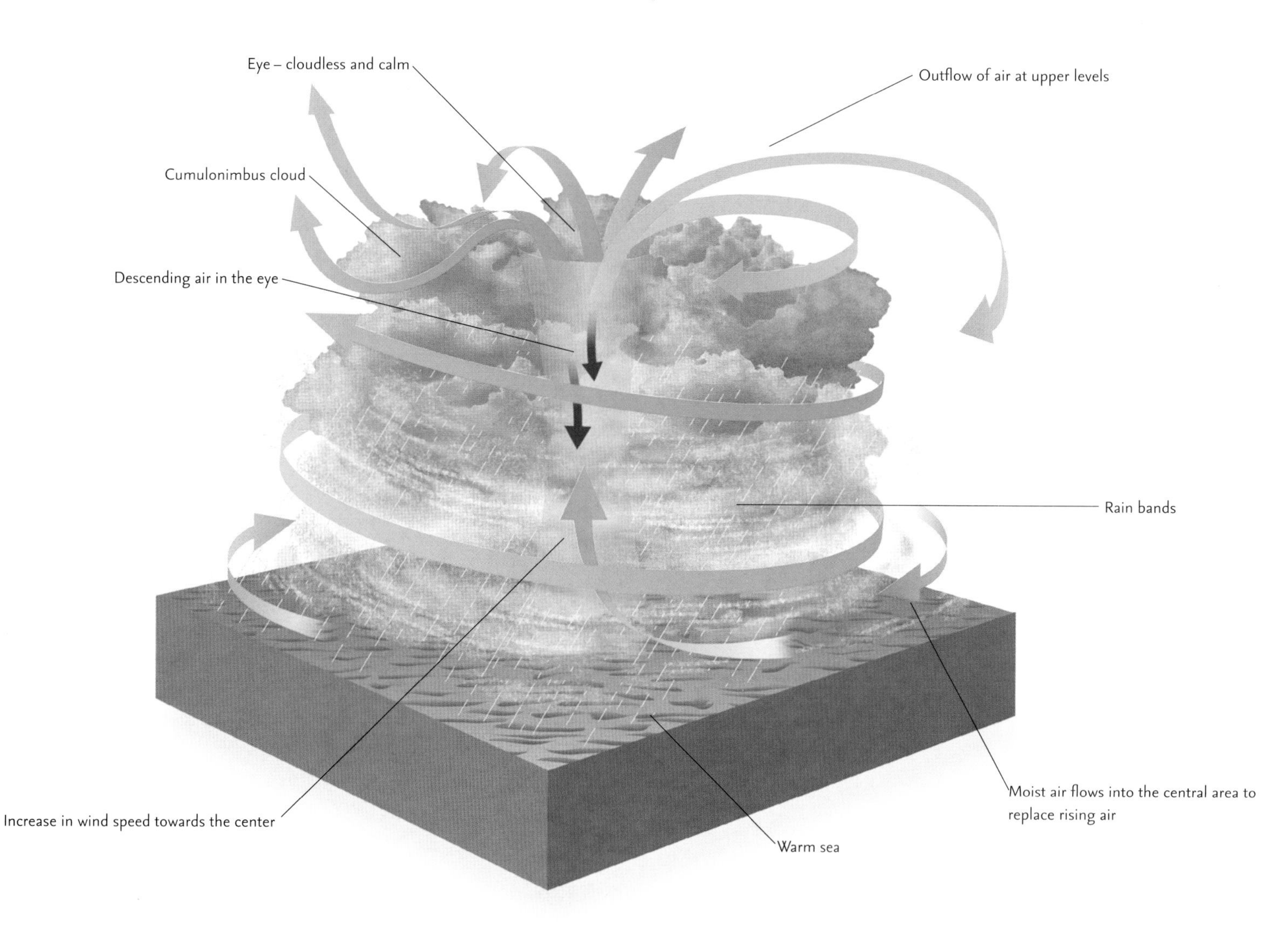

Observing and forecasting tropical cyclones

Forecast Centers

The World Meteorological Organization has designated six Regional Specialized Meteorological Centers (RSMCs) in tropical meteorology and five Tropical Cyclone Warning Centers (TCWCs) with regional responsibility for tropical cyclone forecasting. The six RMSCs are the National Hurricane Center in Miami, Florida, the Central Pacific Hurricane Center in Honolulu, Hawaii, the Japan Meteorological Agency in Tokyo, the Indian Meteorological Department in New Delhi, the Fiji Meteorological Service, and Meteo-France's La Reunion Tropical Cyclone Warning Center.

The five TCWCs are the Australian Bureau of Meteorology's centers in Brisbane, Darwin and Perth, the National Weather Service, Papua New Guinea in Port Moresby, and the Meteorological Service of New Zealand in Wellington.

In addition, the Joint Typhoon Warning Center (JTWC), a joint United States Navy–United States Air Force task force located at the Naval Pacific Meteorology and Oceanography Center in Pearl Harbor, Hawaii, has overlapping forecast responsibility for the issuance of tropical cyclone warnings in the north west Pacific Ocean, south Pacific Ocean and Indian Ocean for United States Department of Defense interests. The JTWC provides support to all branches of the U.S. Department of Defense as well as other U.S. government agencies. Their products are intended primarily for the protection of military ships and aircraft as well as military installations jointly operated with other countries around the world.

Observational and Forecasting Tools

Tools for observing and forecasting tropical cyclones have greatly evolved over the past decade with the advent of new technologies, particularly in the areas of improved satellite data, numerical weather prediction for tropical cyclone track forecasts, and aircraft reconnaissance. This section mentions some of the newer tools now in use.

Passive microwave satellite imagery has been available to forecasters over the past several years. Because of their long wavelengths, compared to the visible and infrared, microwaves have special properties which are important for remote sensing. The longer wavelength microwave radiation can penetrate through cloud cover, haze, dust, and all but the heaviest rainfall. This property of microwave

imagery has greatly aided in the detection of low-level cloud features previously hidden by colder upper level clouds. These low-level cloud features can provide clues to locations of tropical cyclone centers. Microwave satellite imagery is currently available on the NASA TRMM and Aqua satellites.

Scatterometer data became routinely available to forecasters for the first time in June 1999 with the launch of QuikScat. However, prior to QuikScat, NSCAT scatterometer data was available for a short time in 1997, and ERS scatterometer data was available as early as 1996. Scatterometers are essentially radars that transmit microwave pulses down to the earth's surface and then measure the power which is scattered back to the instrument. This "backscattered" power is related to surface roughness. Over water, the surface roughness is highly correlated with the near-surface wind speed and direction. Hence, wind speed and direction at a height of 33 ft (10 m) over the ocean surface are retrieved from measurements of the scatterometer's backscattered power. QuikScat has aided forecasters in determining the location of tropical cyclone centers and providing better detail of tropical cyclone wind fields, particularly the radius of tropical storm force winds in the absence of aircraft reconnaissance data.

Over the past few years, forecasters at the National Hurricane Center have utilized ensemble forecasts of tropical cyclone forecast track models and consensus forecasts based on output from several models instead of output from a single model. During the 2005 and 2006 hurricane seasons, the Florida State University super ensemble model and a consensus of four global models, the GUNA were the best performing models in the Atlantic and east Pacific basins.

Aircraft reconnaissance took a leap forward in 1996 with the advent of NOAA's Gulfstream-IV SP, a state-of-the-art, high-altitude research platform. The Gulfstream was designed to improve NOAA's tropical cyclone forecast capability by being able to deploy dropsondes (weather reconnaissance device) from high altitudes over large areas of open ocean, where few observations are available. This improved sampling of the hurricane environment has resulted in overall track forecast improvements over the past several years.

ABOVE:
Squall line over La Digue, Seychelles.

Hurricanes

A hurricane is the term used to describe tropical cyclones with winds exceeding 64 knots (119 km/hr) which occur in the north Atlantic and east and central Pacific basins in the summer and fall seasons. The word *hurricane*, used in the north Atlantic and northeast Pacific, is derived from the name of a native Caribbean Amerindian storm god, *Huracan*, via Spanish *huracán*. (Huracan is also the source of the word *Orcan*, another word for the European windstorm.)

The hurricane season in the east Pacific basin begins on May 15 and runs through November 30. The Atlantic hurricane season begins two weeks later, June 1, and also ends on November 30. The peak of the Atlantic hurricane season extends from 1 August through the middle of October with a climatological peak around 10 September. However, there have been rare instances where tropical cyclone activity was observed outside these dates. During the record-breaking 2005 Atlantic hurricane season, tropical cyclone activity was observed into January of 2006.

In the Atlantic basin during the period 1950–2005 there were on average ten named tropical storms per year with six of those becoming hurricanes. However, since 1995 there has been an upswing in overall activity with an average of nearly fifteen named storms per year of which eight became hurricanes and four became major hurricanes.

In the Atlantic basin, hurricanes are ranked according to their maximum sustained winds using the Saffir-Simpson scale (see table below) which ranks hurricanes from Category One through Category Five. A Category Three or greater hurricane is considered a major hurricane.

As noted in the previous section, the National Hurricane Center in Miami, Florida is the RSMC which has forecast responsibility for the Atlantic and northeast Pacific (east of 140°W) basins. The Central Pacific Hurricane Center, co-located with the Weather Forecast Office in Honolulu, has forecast responsibility for the north central Pacific between 140°W and the International Date Line (180°W).

Tropical cyclone activity in the Atlantic basin shattered all previous records during the 2005 season with twenty-eight tropical storms of which fifteen became hurricanes and seven major hurricanes. Hurricane Katrina was the costliest hurricane ever to hit the U.S. with insured and non-insured damages estimated at $81 billion. In addition Katrina proved to be the deadliest hurricane to strike the U.S. since the Palm Beach-Lake Okeechobee hurricane of September 1928, killing more than 1,500 persons. This is the known death toll directly caused by the storm as of 2007 and there is a chance this could be revised downward. Most of the deaths in Katrina were due to the massive storm surge from the hurricane, which reached 24–28 ft (7.3–8.5 m) above normal in the vicinity of St Louis Bay, Mississippi. The city of New Orleans and the Mississippi coast are still recovering from the effects of Katrina and will likely take many years before returning to normal.

RIGHT:
2115 UTC August 28, 2005 visible satellite image of Hurricane Katrina at peak Category 5 intensity. The hurricane was located about 250 miles (400 km) south-southeast of New Orleans, Louisiana. Estimated central pressure was 902 millibars with winds of 175 mi/hr (283 km/hr).

Saffir-Simpson Scale

Category	Maximum sustained winds	Effects
One	Winds 74–95 mi/hr 64–82 knots 119–153 km/hr Storm surge approx 4–5 ft (1–1.5 m)	No real damage to building structures. Damage primarily to unanchored mobile homes, shrubbery, and trees. Some damage to poorly constructed signs. Also, some coastal road flooding and minor pier damage. Hurricane Gaston of 2004 was a Category One hurricane which made landfall along the central South Carolina coast.
Two	Winds 96–110 mi/hr 83–95 knots 154–177 km/hr Storm surge approx 6–8 ft (1.8–2.5 m)	Some roofing material, door, and window damage of buildings. Considerable damage to shrubbery and trees with some trees blown down. Considerable damage to mobile homes, poorly constructed signs, and piers. Coastal and low-lying escape routes flood 2–4 hours before arrival of the hurricane center. Small craft in unprotected anchorages break moorings. Hurricane Isabel of 2003 made landfall near Drum Inlet on the Outer Banks of North Carolina as a Category Two hurricane.
Three	Winds 111–130 mi/hr 96–113 knots 178–209 km/hr Storm surge approx 9–12 ft (2.7–3.6 m)	Some structural damage to small residences and utility buildings with a minor amount of curtainwall failures. Damage to shrubbery and trees with foliage blown off trees and large trees blown down. Mobile homes and poorly constructed signs are destroyed. Low-lying escape routes are cut by rising water 3–5 hours before arrival of the center of the hurricane. Flooding near the coast destroys smaller structures with larger structures damaged by battering from floating debris. Terrain continuously lower than 5 ft above mean sea level may be flooded inland 8 miles (13 km) or more. Evacuation of low-lying residences with several blocks of the shoreline may be required. Hurricanes Katrina and Rita of 2005 were Category Three hurricanes when they made landfall in Louisiana.
Four	Winds 131–155 mi/hr 114–135 knots 210–249 km/hr Storm surge approx 13–17 ft (4–5 m)	More extensive curtainwall failures with some complete roof structure failures on small residences. Shrubs, trees, and all signs are blown down. Complete destruction of mobile homes. Extensive damage to doors and windows. Low-lying escape routes may be cut by rising water 3–5 hours before arrival of the center of the hurricane. Major damage to lower floors of structures near the shore. Terrain lower than 10 ft above sea level may be flooded requiring massive evacuation of residential areas as far inland as 6 miles (10 km). Hurricane Charley of 2004 was a Category Four hurricane that made landfall in Charlotte County, Florida with winds of 150 mi/hr.
Five	Winds >156 mi/hr >135 knots >250 km/hr Storm surge approx 18 ft (5.5 m)	Complete roof failure on many residences and industrial buildings. Some complete building failures with small utility buildings blown over or away. All shrubs, trees, and signs blown down. Complete destruction of mobile homes. Severe and extensive window and door damage. Low-lying escape routes are cut by rising water 3–5 hours before arrival of the center of the hurricane. Major damage to lower floors of all structures located less than 15 ft above sea level and within 500 yards of the shoreline. Massive evacuation of residential areas on low ground within 5–10 miles (8–16 km) of the shoreline may be required. Only 3 Category Five Hurricanes have made landfall in the United States since records began: The Labor Day Hurricane of 1935, Hurricane Camille (1969), and Hurricane Andrew in August, 1992.

RIGHT:
Hurricane Kenna was the strongest storm to threaten the Americas in 2002. Kenna was born in the warm tropical waters of the eastern Pacific south of Mexico. This image shows the rain structure inside the rainbands and inner core of Kenna. Red and yellow colors indicate the most intense rains. The compact eye is well formed and is flanked by towering thunderstorm clouds. These towers, which are 10–10.5 miles (16–17 km) tall, contain the heaviest rains and act to energize the core of the storm, sustaining winds of nearly 140 mi/hr (225 km/hr).

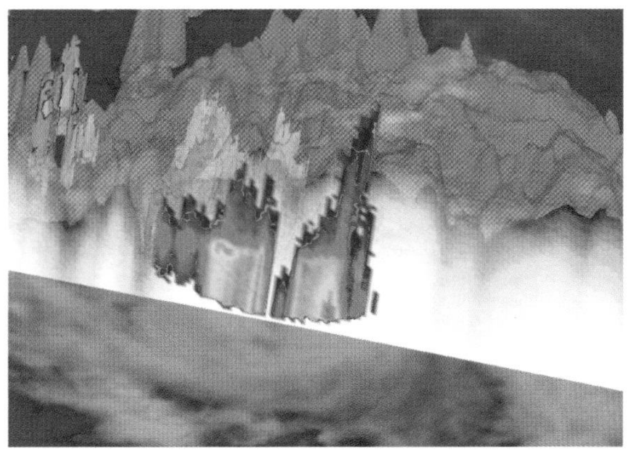

RIGHT:
The maximum sustained winds at the time of this image of Hurricane Andrew were 165 mi/hr (265 km/hr). Andrew was re-analysed as a Category 5 landfall for South Florida. The image was taken very near the time of landfall.

ABOVE:
Inundated area in New Orleans, Louisiana, U.S.A. on September 11, 2005 following the breaking of the levees surrounding the city as the result of Hurricane Katrina.

BELOW:
A sequence of satellite images showing the track of Hurricane Ivan from September 5 to September 15, 2004. It approached the Caribbean from the western Atlantic Ocean and then moved through the Caribbean Sea to the south of Puerto Rico, Haiti, Jamaica, and Cuba. On reaching the Yucatan Channel it headed north, passing over western Florida, U.S.A..

ABOVE:
Large menhaden fishing boats swept onto the highway in Plaquemines Parish, Louisiana, U.S.A. as a result of Hurricane Katrina.

Typhoons

ABOVE:
Three different typhoons spinning over the western Pacific Ocean on August 7, 2006. The strongest of the three, Typhoon Saomai (lower right), had sustained winds of around 85 mi/hr (140 km/hr) and was predicted to gather strength before coming ashore in China.

A typhoon is the term used to describe a tropical cyclone with winds exceeding 64 knots (119 km/hr) occurring in the western north Pacific basin. The word typhoon, used today in the northwest Pacific, has two possible and equally plausible origins. The first is from the Chinese (Cantonese: "daaih fūng"; Mandarin: "dà fēng") which means "great wind." The Chinese term as "táifēng," and "taifū" in Japanese, has an independent origin traceable variously to hongthai, going back to the Song (960-1278) and Yuan (1260-1341) dynasties. Alternatively, the word may be derived from Urdu, Persian and Arabic "ţūfān," which in turn originates from Greek "tuphōn," a monster in Greek mythology responsible for hot winds. The related Portuguese word "tufão," used in Portuguese for any tropical cyclone, is also derived from Greek "tuphōn." There are local terms for tropical cyclones. In the Philippines tropical cyclones are called "Bagyo" or "Baguio." However these names are not used in operational warnings by the various tropical cyclone warning centers.

The typhoon season in the western north Pacific basin begins in April and ends in January of the following year. The western south Pacific typhoon season begins in October and continues until the following May. In an average season, twenty-six tropical cyclones form in the northwest Pacific with seventeen becoming typhoons and nine becoming the equivalent of a major hurricane in the Atlantic basin with winds exceeding 100 knots (185 km/hr). Once a system attains typhoon status, the only classification that exists is between typhoons and super typhoons, a designation used only by JTWC. Super typhoons have winds in excess of 130 knots (240 km/hr), equivalent to a strong Category Four hurricane in the Atlantic basin.

The Japanese Meteorological Agency RSMC has the responsibility for issuing typhoon advisories over the northwest Pacific, west of the International Date Line to 100°E. However as noted above, the JTWC has overlapping forecast responsibility in this area as well as the south Pacific and Indian Oceans.

The 2006 typhoon season was very active with several late-season typhoons in succession striking the Philippines. The first typhoon, Cimaron, struck the island of Luzon on October 29 as a super typhoon with 140-knot winds. Typhoon Cheba, following on the heels of Cimaron, developed rapidly into a super typhoon just east of Luzon on November 10, but fortunately weakened before landfall the next day in Casiguran, Aurora in Luzon. A third typhoon, Durian, reached super typhoon status on November 29 and made landfall in Cantanduanes and Albay province just south of Luzon on November 30. A fourth typhoon, Utor, made landfall in the central Philippines on December 9. The four typhoons were responsible for several thousand deaths in the Philippines.

Cyclones

A cyclone is the term used to describe a tropical cyclone occurring over the southwestern Pacific and southern Indian Oceans near Australia, India and the southeast coast of Africa. The word cyclone was coined by a Captain Henry Piddington, who used it to refer to the storm that blew a sailing ship in circles off Mauritius in February of 1845.

The cyclone season extends from October to the following May with a peak in February and March. In an average season there are eleven tropical cyclones which develop in the Australian Bureau of Meteorology's area of responsibility, of which five reach Category Two or the equivalent of hurricane intensity and two reach Category Three or the equivalent of major hurricane intensity in the Atlantic basin. In the south Indian Ocean, an average of twenty tropical cyclones develop, of which ten reach the equivalent of minimal hurricane intensity and four reach the equivalent of major hurricane intensity during the period October through May.

There are several forecast centers responsible for forecasting cyclones. Over the Pacific Ocean, the Fiji RSMC (Regional Specialized Meteorological Center) on the island of Nadi issues warnings for the south Pacific to 25°S between 120°W and 160°E. The remainder of the Pacific Ocean between 120°W and 160°E is covered by the TCWC (Tropical Cyclone Warning Center) in Wellington, New Zealand.

The TCWCs in Brisbane, Darwin and Perth, which comprise the Australian Bureau of Meteorology, as well as the TCWC in Port Moresby have forecast responsibility for the remainder of the Pacific Ocean and the far-eastern Indian Ocean east of 88°E.

The Australian Bureau of Meteorology uses a one to five scale called "tropical cyclone severity categories." Unlike the Saffir-Simpson Hurricane Scale, severity categories are based on estimated maximum wind gusts. In general, a Category One storm features gusts less than 67 knots (125 km/hr) while gusts in a Category Five cyclone are at least 150 knots (280 km/hr). Category Three, Four, and Five cyclones are classified as "severe."

Cyclone forecast responsibility for the north Indian Ocean between 100°E and the Arabian coast is covered by the Indian Meteorological Department RSMC. Meteo-France's La Reunion Tropical Cyclone Warning Center forecasts for the south Indian Ocean to 30°S west of 88°E to the east coast of Africa and the island of Madagascar, utilizing a different intensity scale and terminology for storm classification than the Australian Bureau of Meteorology. The terms used in the southwestern Indian Ocean include the following: *Moderate Tropical Storm* means a tropical disturbance in which the maximum of the average wind speed is 39–55 mi/hr/34–47 knots (63–88 km/hr).

Severe Tropical Storm indicates a tropical disturbance in which the maximum of the average wind speed is 55–73 mi/hr/48–63 knots (89–118 km/hr).

Tropical Cyclone is used for a tropical disturbance in which the maximum of the average wind speed is 74–103 mi/hr/64–89 knots (119–165 km/hr).

The term *Intense Tropical Cyclone* labels tropical disturbances in which the maximum of the average wind speed is 103–132 mi/hr/90–115 knots (166–212 km/hr).

Very Intense Tropical Cyclone designates a tropical disturbance in which the maximum of the average wind speed is greater than 132 mi/hr/115 knots (212 km/hr).

Cyclone Intensity Scale

Category	Highest gusts	Effects
One	<77 mi/hr <67 knots < 125 km/hr GALES	Damage to crops, trees and caravans. Small craft may drag their moorings. Negligible house damage.
Two	77–105 mi/hr 67–91 knots 125–169 km/hr DESTRUCTIVE WINDS	Significant damage to signs, trees and caravans. Heavy damage to crops. Risk of power failure. Small craft break moorings. Minor house damage.
Three	106–139 mi/hr 92–120 knots 170–224 km/hr VERY DESTRUCTIVE WINDS	Some roof and structural damage to houses. Some caravans destroyed. Power failures likely.
Four	140–173 mi/hr 121–150 knots 225–279 km/hr VERY DESTRUCTIVE WINDS	Significant roofing loss and structural damage. Many caravans destroyed or blown away. Dangerous airborne debris. Widespread power outages.
Five	>174 mi/hr >151 knots > 280 km/hr VERY DESTRUCTIVE WINDS	Extremely dangerous with widespread destruction.

BELOW:
Palm trees blowing on a stormy day in Mumbai (Bombay). In 2005 Mumbai experienced it's worst flooding in living memory. The floods were exacerbated due to the removal of large areas of mangrove swamps which could have absorbed some of the floodwater.

Monsoons

The term monsoon is used to describe prevailing seasonal winds. The term was originally applied to the winds over the Arabian Sea which blow from the northeast during the late fall, winter and early spring of the year and from the southwest during the late spring, summer, and early fall of the year. The term is actually derived from the Arabic *mausim*, meaning season. However the term has been extended to other areas of the world which experience prevailing winds on a seasonal basis.

Monsoons are primarily driven by greater annual variations in temperatures over land masses than over the oceans. This results in higher pressure over the continents in winter and lower pressure in the summer months. Monsoons are the most prevalent on the southern and eastern portions of the continent of Asia, which has the largest landmass. However, monsoons occur in other areas of the globe, including Spain, northern Australia, Africa (with the exception of the Mediterranean), the southwestern U.S., and the west coast of Chile. In India the term is applied to the southwest monsoon which brings the summer rains and begins the wet season.

The height of the land plays an important role in enhancing the monsoon, particularly over southern Asia where warm moist air flowing up from the Indian Ocean rises up the Himalayan Mountains to enhance the rainfall over the Indian subcontinent.

The lower sea-level pressures within a monsoon region comprise what is known as the monsoon trough. Within this trough, cyclonic disturbances, also known as monsoon depressions, can develop. The term originally referred to weak cyclonic disturbances which develop over the Bay of Bengal and move northwestward over India. The term also describes disturbances forming within the monsoon trough over Australia and western north Pacific. The term monsoon depression has also been used to describe large cyclonic vortices in the tropical north Pacific which have a diameter in the order of 600 nautical miles (1,000 km) and are characterized by a large inner core of light winds within an area of loosely organized convection. These systems can develop into tropical cyclones if they persist over warm waters.

BELOW LEFT:
Flooded street during heavy monsoon rain in Varanasi (Benares), India.

BELOW:
Monsoon flooding on the plains of India, showing half submerged field systems.

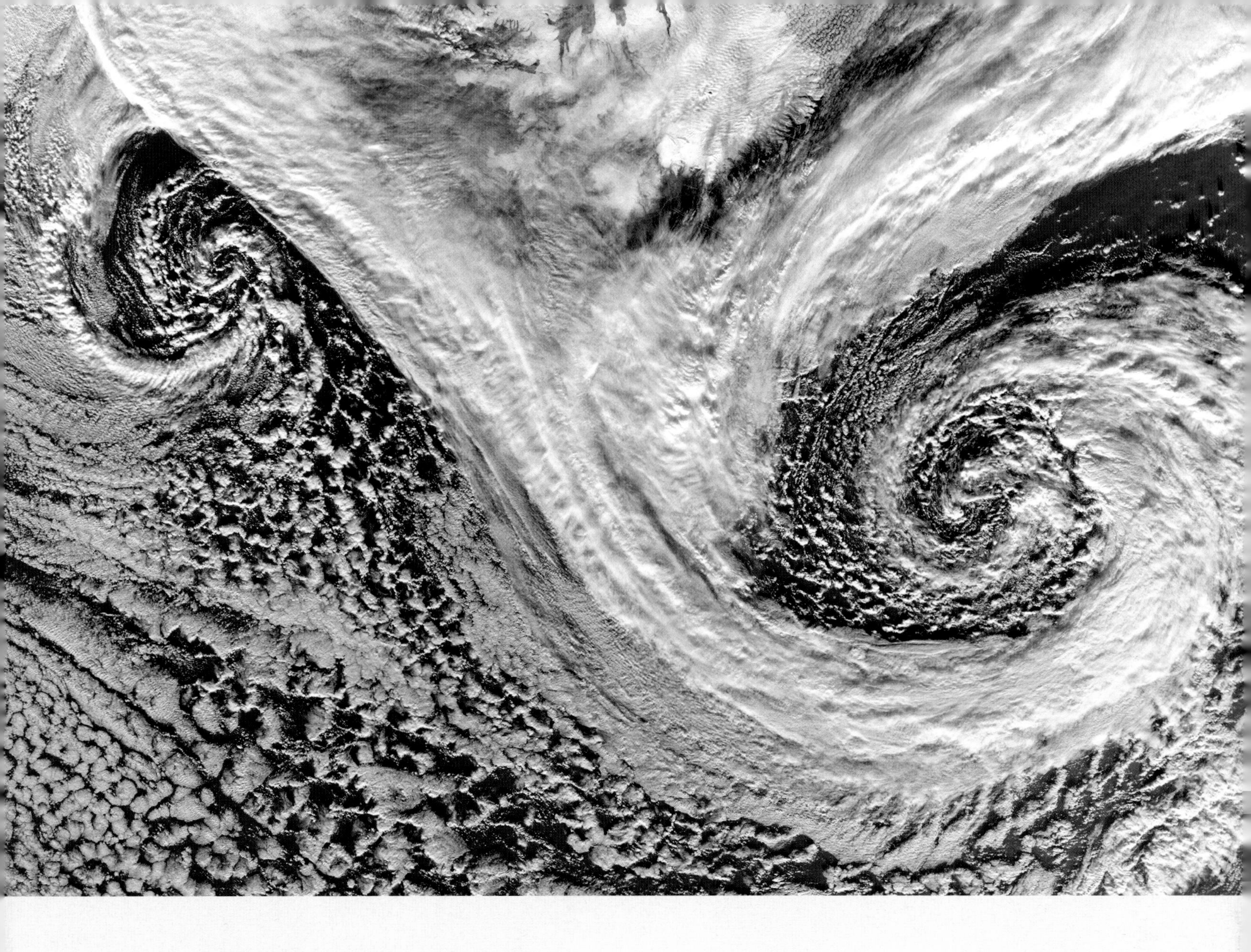

Extra-tropical cyclones

The large oceanic, mid-latitude areas of the world such as the north Pacific, north Atlantic, and 40° latitude area of the southern hemisphere are breeding grounds for enormous oceanic storms. These storms, known as extra-tropical cyclones, can span thousands of miles, with cold fronts that stretch from the Tropics to the Arctic Circle. They transport significant amounts of moisture northward and are the driving force for most of the significant weather which occurs in mid-latitudes during winter.

Since the atmosphere attempts to remain in equilibrium, air will move to equalize differences in temperature, pressure, and moisture. When a warm, moist, lower-pressure tropical air mass encounters a cold, dry, higher-pressure polar air mass, a frontal boundary forms, giving rise to clouds and precipitation. Over time and distances of hundreds of miles, the rotation of the earth and friction cause the mixing air to spin, creating low pressure which spins counter-clockwise in the northern hemisphere and clockwise in the southern hemisphere. The greater the differences the air masses have, the stronger the wind blows and the more intense the cyclone becomes. In this way, massive extra-tropical storms are born.

These great storms are capable of hurricane force winds reaching over 100 mi/hr (160 km/hr) as they impact coasts. They cause blowing spray which greatly reduces visibility at sea and freezing spray which deposits ice on the superstructure of vessels which can cause them to capsize. They spawn thunderstorms with hail, waterspouts, torrential rainfall, and blinding heavy snow. Parades of extra-tropical cyclones build large ocean swells surging to more than 50 ft (15 m). These enormous waves damage large ships and erode coastlines. These storms present a major hazard to transoceanic shipping and fishing and can bring significant precipitation events causing catastrophic flooding and severe weather as they move inland.

Extra-tropical fronts are sometimes referred to as "conveyor belts" since they transport massive amounts of moisture from the Tropics northward. One such conveyor belt is the Pineapple Express, which transports moisture from near Hawaii northeastward, causing prolonged precipitation which can lead to flooding along the U.S. west coast.

ABOVE:
Extra-tropical cyclones near Iceland. There is unusual cloud formation with the double eye of the storm.

Sea fog

Extensive areas of sea fog form over the world's oceans and can cover hundreds of square miles. This fog can become dense, causing visibilities to fall to zero over the open ocean and along coasts. It can persist for several days to even weeks. Sea fog is most common in the summer months, generally along the west coast of continents and at high latitudes, where cold water is most prevalent.

Sea fog is a type of advection fog which forms when air from a warmer water area pushes into a colder water area, causing it to cool and condense, forming a cloud along and near the ocean surface.

In mid-latitudes, such as along the west coast of the U.S. during the dry summer season, air over the continents and above the ocean warms faster than the ocean water. This warmer air acts as a lid, known as an inversion, capping upward vertical movement of colder, heavier, moist oceanic air. Near the ocean surface, air from warmer water areas hundreds of miles offshore often gets pushed into colder, upwelling generated air near the coast, resulting in the formation of dense sea fog. In near-coastal areas, known as fog belts, sea breezes push sea fog inland, providing cooler, moist air to fog- loving trees such as the northern California coastal redwoods and Sitka spruce of the Pacific northwest. Occasionally, this oceanic air surges over a hundred miles inland, bringing a rapid cool-down and increase in moisture to seasonally arid forests, decreasing fire danger.

Sea fog can also make navigation difficult and dangerous for transoceanic shipping and fishing by obstructing the view of rocky coastlines, beaches, breaking waves, coral reefs, lighthouses, and floating objects such as icebergs, trees, and oceangoing vessels.

BELOW:
View of the Golden Gate Bridge in fog, with the city of San Francisco, California, seen in the background.

Sea smoke

When very cold, polar or arctic air pushes into a relatively warm ocean area such as the Gulf Stream, sea smoke, also known as steam fog, forms. It can also form over the ocean when air from a nearby landmass cools substantially on clear, calm nights and then sinks down and outward over the ocean surface. On these clear, calm nights, ocean water remains relatively very warm in comparison to the colder and drier air spreading over it from the land because the heat capacity of water is much higher than that of air. Since cold air sinks because it is heavier than warmer air, as it spreads out over the relatively warm and moist air rising from the ocean surface, it cools the ocean air to its saturation point. Tendrils of sea smoke then rise from the ocean surface.

While steam fog can form anywhere when relatively cold air moves over warm water, it is most common in the autumn and winter over warm ocean currents most prevalent on the eastern side of continents. In these areas, very cold and dry polar and arctic air generated from snow and ice covered land areas pushes out and over the relatively warm and moist ocean.

Steam fog is both beautiful and hazardous, as the rising steam is quite picturesque. At high latitudes, however, this steam fog can hide hull-splitting icebergs and, in below-freezing conditions, can freeze to the decks and superstructures of ships causing slippery conditions.

ABOVE:
Sea smoke, also known as steam fog, rises from the ocean surface much like steam does from a boiling pot of water. It can significantly reduce visibility along and within tens of feet of the ocean surface even with the sun shining from above. This picture shows short-beaked common dolphins in the Atlantic.

Sea breezes

BELOW:
Sea breezes are put to good use by this windsurfer.

Along the coasts of the world, temperature differences between the ocean and the land can result in onshore winds known as sea breezes. Sea breezes bring cooler air inland along and near the coast as marine air surges inland to replace hotter air rising from the land surface. They can also initiate rainfall.

Sea breezes are most prevalent in the Tropics and mid-latitudes on hot summer days. When the sun heats the air near the surface of the earth to a temperature warmer than that of the air above it, it rises. Since the land mass warms faster than the ocean, there is more rising motion over the land compared to over the water. As hot air rises over the land, cooler ocean air moves in to replace it. The greater the difference in temperature between the ocean and the land, the stronger the sea breeze blows and the farther inland it goes.

Coastlines extending for hundreds of miles develop sea breezes which are so pronounced that they sometimes resemble cold fronts. These "sea breeze fronts" can initiate showers and thunderstorms along and near the coast, where air on the leading edge of the sea breeze mixes with the rising heat from the land.

They can bring relief from summer heat, but can also result in extremely high humidity, such as occurs in the eastern U.S. in the summer. Occasionally, sea breeze fronts interact with thunderstorms initiated by cold fronts coming from a direction other than from the ocean. As these thunderstorms merge with the sea breeze front, development of severe or even tornadic supercell thunderstorms can occur.

Dynamic coasts

Dynamic coasts
Introduction

To understand geological processes, sometimes one must think in time scales not encountered during everyday experiences. Mountains build up and cliffs erode, not in days or months, but in kilo-annum (one thousand years) and in mega-annum (one million years). Geologists refer to time scales like these as "geologic time."

If a timeline of Earth's history started at the Golden Gate Bridge in California and stretched all the way to the Lincoln Memorial in Washington, D.C., the first anatomically modern humans in the fossil record would not appear until the last 125 yards (115 m). The time since the last glacial advance would be at the entrance of the Memorial, and the Industrial Revolution would have started 4 inches (10 cm) from Lincoln's feet.

The Earth's appearance has been constantly changing throughout its extremely long history. One of the most dynamic environments is the coastal zone. Here, at the land/ocean interface, waves and the tidal currents work at the terrain all day, every day. Some of the ocean's handiwork is apparent to a human observer over a few months or years, while other processes take centuries to have a noticeable effect.

Beaches and barrier islands can change in one day due to disastrous events or they can change over a season due to moderate repetitive forces. A strong coastal storm, such as a hurricane, will produce waves and large tides which rework sand, altering the shape and size of an island overnight. Conversely, during the summer, a few months of gentle waves move sand landward from offshore sandbars, replenishing the beach.

Surtsey, an island off the southern coast of Iceland, literally formed overnight. In November 1963, fishermen saw a plume of smoke on the horizon. They decided to investigate, thinking that another fishing vessel was in distress and needed immediate assistance. Upon arrival, the fishermen discovered a billowing volcano where the nautical chart indicated open ocean. Scientists carefully began to document the birth of the island until the eruption stopped four years later. In the forty years since then, biologists have watched the island's environment change from a barren rock field to a vibrant ecosystem of plants, insects, birds, and marine mammals.

Coastline evolution can also occur over millennia. The beautiful coastline of southeastern England is the product of waves and storms combined with sea level change and crustal uplift. One hundred million years ago, a shallow sea covered the land and the shells of marine creatures accumulated, forming a thick layer of earth. As sea level fell, the waves and tides relentlessly hammered the newly formed layer of earth for millions of years, eventually forming the English Channel. Part of the resulting coastline, known today as the White Cliffs of Dover, is one of England's most popular tourist destinations and a natural treasure.

Barrier islands will continually shift with the changing seas and river deltas will grow with each flood. Given a few years, humans can easily destroy what took nature thousands and millions of years to create. Dams, jetties, and other man-made structures stop many natural cycles of creation and destruction. Cities often drain marshlands to expand their farmland and accommodate population growth. In the past, we humans have seen swamps and sand, not vibrant ecosystems perfected over time.

An advance in our understanding of the dynamic coastline is slowly changing the attitudes of businesses, communities, and governments. People now understand that drawing lines on a map cannot stop natural geologic cycles. Geologic time, once an abstract concept nurtured by a handful of scholars, is finally crossing into the mainstream.

RIGHT:
Surtsey in November 1963 (top) and thirty-seven years later in 2000 (bottom). An undersea eruption off the coast of Iceland created the new island. Over a period of four years Surtsey erupted and lava flowed. An island was eventually created to a height of 554 ft (169 m) above sea level with an area of 0.97 sq miles (2.5 sq km).

PREVIOUS PAGE:
Ocean waves hitting a rocky coastline.

Beaches and barrier islands

Beaches and barrier islands are constantly moving and changing. Wave action, tides, and ocean currents transport the sand and shell fragments composing these areas. Migrating down current over time, their movement completes a cycle of destruction and replenishment.

Waves approach the beach from many different directions. When they break on the sand, the indirect angles add up to form a current which runs along the shore. This longshore current moves sand from one location to another. Sediment follows a zigzag pattern down the beach along the path of the breaking waves.

Seasonal changes in the weather move sand either on or off shore. Powerful fall and winter storms pull sediment offshore to form sandbars. The calm seas and gentle waves during the spring and summer move the sand from the sandbars to the beach.

Entire barrier islands migrate with the currents. The longshore current slowly strips away the up-current beach and deposits it further down the shore. The sediment at the down-current tip of the barrier island moves offshore and eventually settles on the next island. These combined movements result in the net downstream movement of the whole island.

Dunes are the build-up of sediments upon beaches and islands. Vegetation captures sand blown around by the wind and mounds begin to form. These mounds grow and merge forming large sand dunes. Dunes are composed of finer sediments on the slope facing away from the beach since the finer particles fall out of the air in the dune's wind shadow. The erection of buildings on barrier islands blocks the natural migration of sand and damages the natural dune system.

In many areas, jetties and other artificial structures interrupt the normal movement of sediment by the longshore current. The sand no longer replenishes the barrier islands and sand dunes, causing them to shrink or even disappear. Structures also slow the water moving in and out of inlets between the islands. The slower water drops the sediment and people must constantly dredge these inlets to keep them safe for navigation by boats.

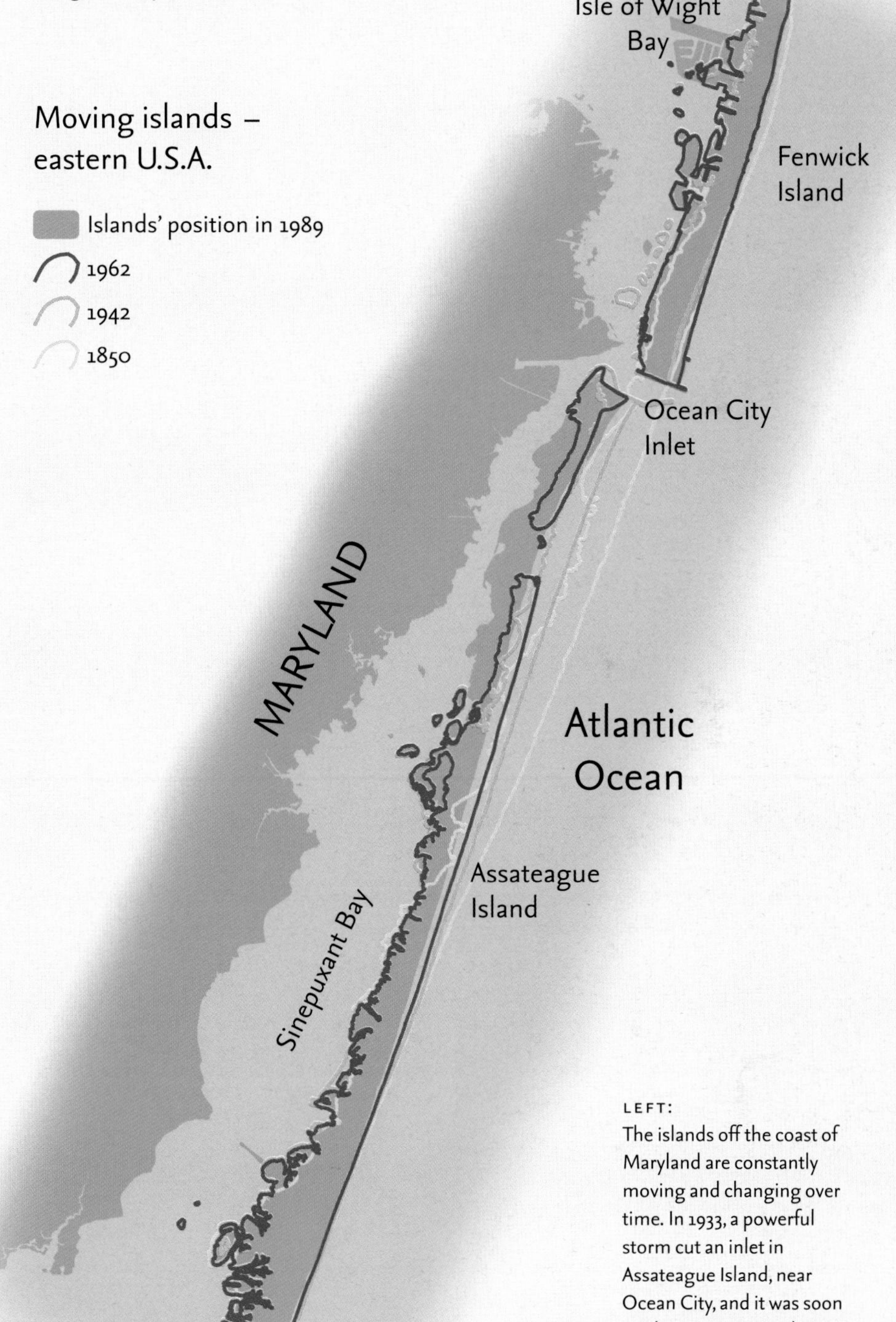

Moving islands – eastern U.S.A.

Islands' position in 1989
1962
1942
1850

Isle of Wight Bay
Fenwick Island
Ocean City Inlet
MARYLAND
Atlantic Ocean
Assateague Island
Sinepuxant Bay

LEFT:
Sporadic reworking of the shoreline by the ocean is evident from these pictures taken before and after Hurricane Fran made land fall at North Topsail Island, North Carolina on September 5, 1996.

LEFT:
The islands off the coast of Maryland are constantly moving and changing over time. In 1933, a powerful storm cut an inlet in Assateague Island, near Ocean City, and it was soon made permanent with jetties on either side. The beach just south of the jetty no longer received sand and began to erode. The beaches furthest to the south on Assateague Island received sand which bypassed the jetties and did not erode.

Rocky shores

The mention of a rocky coastline evokes images of raw beauty: rugged cliff faces, swirling foamy water, and waves throwing spray across the landscape. Artists often use it as a dramatic metaphor for permanence in a chaotic world. The terrain appears unvarying during a human lifetime, but over long periods, rocky coasts are a very dynamic environment.

Every day, waves containing gravel, wood, and other objects chisel away at the rocks. The seawater itself can dissolve and deteriorate the minerals in the cliff face. Altogether, over hundreds and thousands of years, these processes carve a notch in the base of the cliff. After more erosion, the notch becomes a cave, which continues to grow. As the cave grows, the foundation of the cliff cannot support the weight of the rock above it. Eventually, the entire cliff face crumbles into the ocean, forming a large rock pile that protects a newly-exposed cliff. Over hundreds and thousands of years, the ocean erodes and transports the rock pile down the shore until the new cliff face is exposed to the elements and the cycle begins again.

The cliff face above the ocean also erodes over thousands of years. The remaining rock below the ocean surface forms a large, flat platform of rock

exposed at low tide called a marine terrace. In some areas, the marine terrace becomes a raised beach when sea level lowers relative to the platform or when tectonic activities, such as earthquakes, raise the coastline. Raised beaches also form when the land rises due to the loss of weight (called isostatic adjustment) caused by melting of glaciers.

Sometimes the erosion does not happen equally along the coast because of variations on the sea floor or because previously eroded cliffs magnify the waves toward certain areas. In addition, cliff faces are not always uniform and can be composed of several different minerals which erode at different rates.

A stack is a rock island shaped like a pole which forms when all the rock around it erodes. Blowholes are openings in the ceiling of sea caves. When a wave breaks into the cave, air pressure forces water up through the blowhole. This results in a spectacular spray of mist across the rocks which resembles the blow a whale makes when it surfaces to breathe.

ABOVE:
The action of waves over countless millennia has shaped the hard granite at Cape Leeuwin, Australia into smooth beds.

ABOVE:
Volcanic activity in the Katmai area of the Aleutian Range, southwest Alaska.

Volcanic islands

Thousands of islands around the world are products of volcanic eruptions. Sometimes the slow seeping of magma out of the earth's crust over hundreds of thousands of years will build a large island. At other times violent explosions and large flows will form an island over a few years. Scientists classify the three main types of volcanic islands based on their method of formation. These islands form all over the world, but most are located in the Pacific Ocean, earning this region the nickname "Ring of Fire."

Volcanic island arcs are the largest volcanic islands. They form near a continent when the oceanic plate moves under the continental plate in a process called subduction. This movement causes the formation of magma which works its way to the earth's surface through weak points in the crust. Subduction happens over large areas along the edges of the plates. The Aleutian Islands in southwest Alaska and the islands of Japan are examples of volcanic island arcs. Earthquakes and volcano eruptions are a constant threat to people living in these island chains.

A hotspot volcanic island forms when a plate moves in one direction over a single point source of magma. The change in magma flow over time causes islands of different sizes to form long chains in the direction of the plate motion. Tuzo Wilson, a Canadian

geophysicist, first recognized the Hawaiian Islands as hotspot volcanic islands in the 1960s. He realized that as the islands progress to the northwest, they become increasingly older until the ocean completely erodes them into underwater mountains called seamounts.

If the water is warm and clear, a coral reef may form around a volcanic island. When the volcano erodes, only the reef remains with a lagoon inside. Soon the reef traps sediment, forming a flat island just above sea level called an atoll. Hundreds of atolls in every ocean are at risk due to the global increase in sea level. Rising sea level may submerge these low-lying islands and force their inhabitants to relocate.

ABOVE:
Three-dimensional bathymetric view of the Maug caldera, Pacific Ocean. It is a large volcanic crater which is now flooded by the sea. Depths in this image range from 790 to 82 feet (241 to 25 m). It is two times vertically exaggerated.

TOP LEFT:
Aerial view of the lava bench at East Lae'apuki, Hawaiian Islands, looking northeastward at 2pm, September 20, 2006. The plume from the East Ka'ili'ili entry is drifting along shore. Silver lava at the top of the sea cliff near the bench (center) marks the location of a brief breakout earlier in the day.

TOP RIGHT:
Breakout from the lava tube just inland of the sea cliff at East Lae'apuki resumed at about 3.30pm. This activity continued on-and-off until daybreak on September 21. Lava streamed over the sea cliff to fall about 49 ft (15 m) onto the bench below.

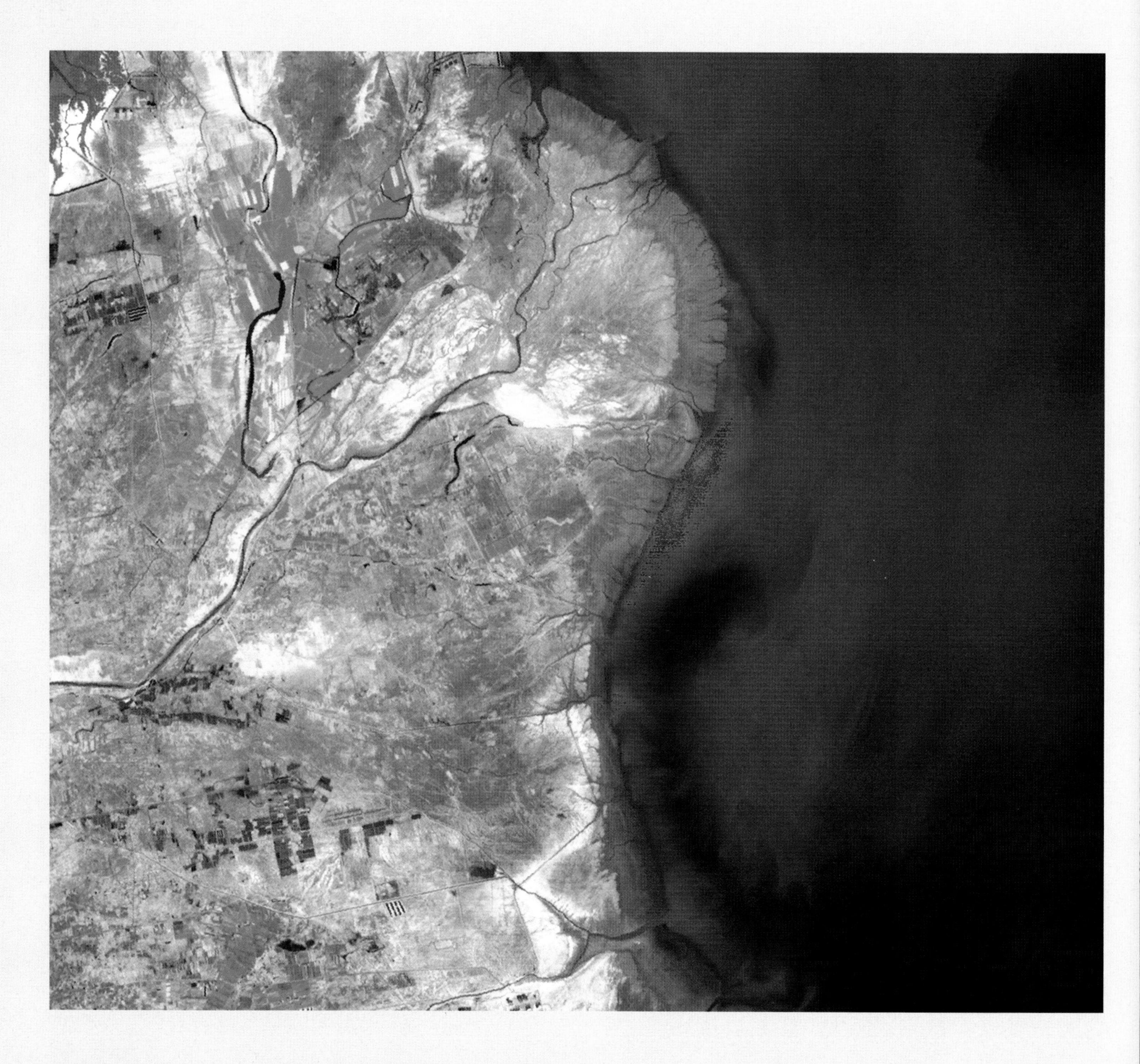

River outlets and deltas

Around 450 BC, the Greek historian Herodotus observed that the shape of the land where the Nile River met the Mediterranean Sea resembled a triangle. He called this feature a "delta" after the angular Greek letter Δ.

Deltas are very common where rivers empty into larger bodies of water, but not all rivers have deltas. They occur at the mouths of rivers where adequate sediment delivery builds up land faster than sea level rise can wash it away.

Land builds up because rivers accumulate sediment on their journey to the sea. The soil characteristics of the watershed and speed of the river determines the types and amounts of sediment carried downstream. Faster water carries larger sediments and slower water carries smaller sediments. When a river slows near the ocean, large sand and mud particles settle to the bottom first. Further downstream, the smaller clay and silt particles slowly drift to the bottom. A major characteristic of deltas is the gradation of larger to smaller sediments as the landform progresses from river mouth to offshore.

Scientists classify deltas according to the behavior of the water at the river/ocean boundary. River-dominated deltas form in areas where tides and waves are small and river discharge is large. The sediment builds up land until, in a process called avulsion, the river changes course upstream resulting in a different path to the sea. Periodic avulsion on a scale of hundreds to thousands of years causes the Mississippi river delta to give the impression of a bird's foot.

Tide-dominated deltas have the deposited river sediment reworked by the tide to form islands and peninsulas oriented in the direction of the currents. The amount of water moved by the tide must be much greater than the amount of water delivered by the

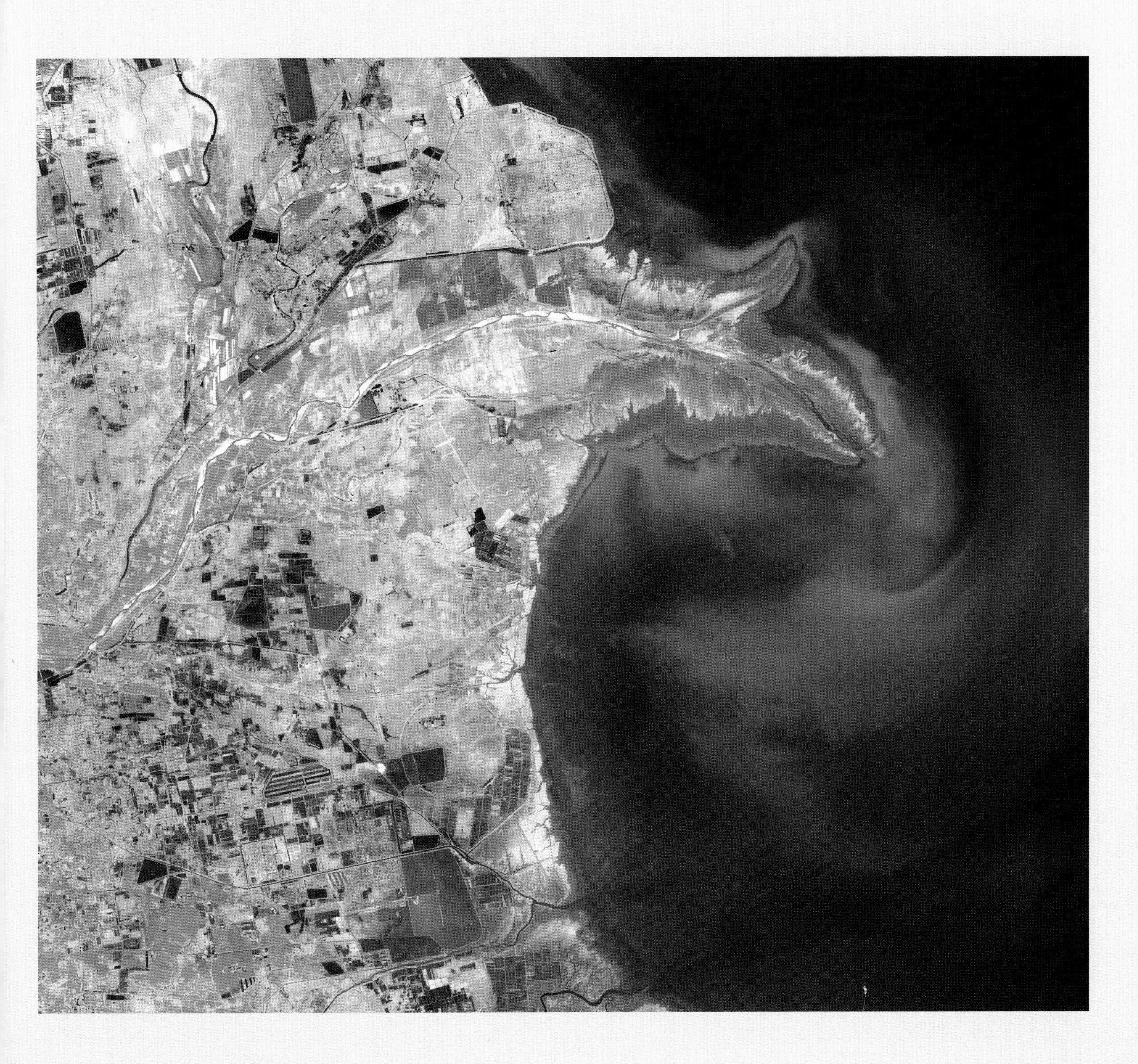

river. Nutrients from both the river and the daily tidal migration make the extensive wetlands of these deltas very productive. The fertile, tide-dominated Ganges-Brahmaputra delta in India provides crops and fish for over 130 million inhabitants.

A wave-dominated delta has large waves and longshore currents which transport the sediment away from the river mouth and along the coast. Usually only large oceans have enough energy to form a wave-dominated delta. In a typical example, storms in the Atlantic Ocean form swells which continually batter the coast of Brazil. The Rio São Francisco drains into the Atlantic and the resulting delta is a small bulge. Down the shore, sediments released by the river become barrier islands, sandbars, or beaches. In addition, the resulting landform is not as fertile as other deltas because the ocean washes away many of the nutrients.

ABOVE:
Yellow River (Huang He), China on May 27, 1979 (left) and after twenty-one years of change on May 2, 2000 (right). As the river travels through north central China it crosses the easily eroded Loess plateau picking up fine-grained deposits. When the river reaches the coast it flows into the sea where the sediments drop out of the current and are deposited into the river delta. Increased demand on the river drastically lowered the water flow in the time between the pictures. Sometimes the river even dried up before reaching the Gulf of Bohai. This greatly reduced the amount of sediment delivered to the delta. A new channel was cut in 1997 which resulted in the growth of the delta to its present configuration.

Natural hazards

Natural hazards
Introduction

People face the threat of death or serious injury from various sea creatures both large and small. This can be as a result of being attacked, stung or poisoned. However, these living or biological hazards are much less of a threat than the powerful physical forces within the oceans. In regions around the globe, individuals in coastal communities are at risk from hurricanes, tsunamis, flooding, storm surges, waterspouts, rip currents, coastal erosion as well as biological hazards. Often, these hazards become disasters because local communities lack the capacity to recognize, understand and prepare for the hazards or to receive information about these conditions and events as they are unfolding. In order to ensure better individual and public safety along the coastlines, the right information, in the right format, must get to the right people at the right time, to inform good decision making. Across the U.S. government and among governments around the world, this is the goal of the Global Earth Observation System of Systems (GEOSS). Among other societal benefits, GEOSS can help reduce the loss of life and property during hazard events and disasters.

The intergovernmental Group on Earth Observations (GEO), which now includes more than seventy countries, the European Commission, and forty-six participating organizations, is leading a worldwide effort to build GEOSS over the next ten years.

GEOSS is a global initiative to achieve comprehensive, coordinated, and sustained observations of the earth in order to meet the need for timely, quality, long-term information as a basis for sound decision making. GEOSS will build on and add value to existing earth observing systems by coordinating efforts, addressing critical gaps, supporting interoperability, sharing information, reaching a common understanding of user requirements, and improving the delivery of information to users.

The major societal benefits of GEOSS include:

•Reducing loss of life and property from natural and human-induced disasters.

•Understanding environmental factors affecting human health and well-being.

•Improving management of energy resources.

•Understanding, assessing, predicting, mitigating, and adapting to climate variability and change.

•Improving water resource management through better understanding of the water cycle.

•Improving weather information, forecasting and warning.

•Improving the management and protection of terrestrial, coastal and marine ecosystems.

•Supporting sustainable agriculture and combating desertification.

•Understanding, monitoring and conserving biodiversity.

Earth observations are integral to reducing losses from hazards, by producing information which is needed throughout the disaster cycle from forecasts and warnings to facilitating pre-event deployment and rapid response. These vital data are also required to support the longer-term processes of mitigation, land-use planning, recovery, and rebuilding.

PREVIOUS PAGE:
Great white shark underwater. This is the largest of the predatory sharks, reaching a length of over six meters. It mainly eats seals, turtles and fish but occasionally attacks humans. Like other sharks, it is able to detect tiny traces of blood in the water and follow them to their source. Photographed off South Africa.

RIGHT:
An artist's impression of many of the elements that make up the Global Earth Observation System.

NOAA
R 330

Tsunamis

Tsunamis are ocean waves produced by earthquakes or underwater landslides. The word is Japanese and means "harbor wave," because of the devastating effects these waves have had on low-lying Japanese coastal communities, although they can occur anywhere. Perhaps the most devastating tsunami in human history was the Indian Ocean tsunami of December 26, 2004, which caused the deaths of over 225,000 people. Tsunamis are often incorrectly referred to as tidal waves, but a tsunami is actually a series of waves that can travel at speeds averaging 450 mi/hr (720 km/hr), reaching up to 600 mi/hr (965 km/hr) in the open ocean.

In the open ocean, tsunamis would not be felt by ships because the wavelength would be hundreds of miles long, with an amplitude of only a few feet. This would also make them unnoticeable from the air. As the waves approach the coast, their speed decreases and their amplitude increases. Unusual wave heights have been known to be over 100 ft (30 m) high. However, waves that are 10–20 ft (3–6 m) high can be very destructive and cause many deaths or injuries.

From an initial tsunami-generating source area, waves travel outward in all directions much like the ripples caused by throwing a rock into a pond. As these waves approach coastal areas, the time between successive wave crests varies from five to ninety minutes. The first wave is usually not the largest in the series of waves, nor is it the most significant. Furthermore, one coastal community may experience no damaging waves while another, not that far away, may experience destructive deadly waves. Depending on a number of factors, some low-lying areas could experience severe inland inundation of water and debris of more than 6 miles (10 km) inland.

RIGHT AND FAR RIGHT:
Lhoknga, Aceh, Indonesia 10 January 2003 before the tsunami. The Indonesian province of Aceh was hit hardest by the earthquake and tsunamis of December 26, 2004. Aceh is located on the northern tip of Sumatra. The death toll in Indonesia reached over 165,000—more than half the global total.

Tsunami aftermath, 29 December 2004. The town of Lhoknga, on the northwest coast of Sumatra near the capital of Aceh, Banda Aceh, was completely destroyed by the tsunami, with the exception of the mosque (white circular feature within the box) in the city's center.

Cross-section of a tsunami

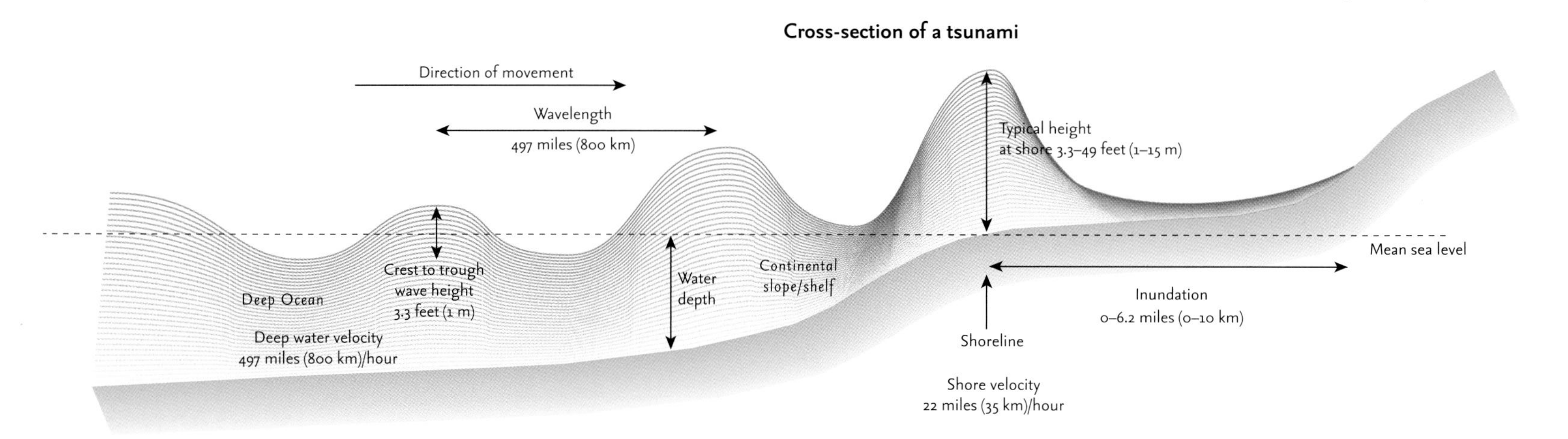

Tsunamis in recent history

Date	Location	Extent	Caused by	Deaths
8 November 1929	Grand Banks of Newfoundland, Canada	Atlantic Ocean	Earthquake of magnitude 7.2 and submarine landslide	29
1 April 1946	Aleutian Islands, Alaska, U.S.A.	Pacific Ocean	Earthquake of magnitude 7.8	165
9 March 1957	Aleutian Islands, Alaska, U.S.A.	Pacific Ocean	Earthquake of magnitude 8.2	-
4 November 1957	Kamchatka Peninsula, Russian Federation	Pacific Ocean	Earthquake of magnitude 8.3	-
22 May 1960	South Central Chile	Pacific Ocean	Earthquake of magnitude 8.6	122
28 March 1964	Prince William Sound, Alaska, U.S.A.	Pacific Ocean	Earthquake of magnitude 8.6	-
29 November 1975	Hawaii, Pacific Ocean	Local	Earthquake of magnitude 7.2	2
26 December 2004	off Sumatra, Indonesia	Indian Ocean	Earthquake of magnitude 9.0	226 408

LEFT:
A tsunami wave approaching Kho Lanta, Thailand December 2004.

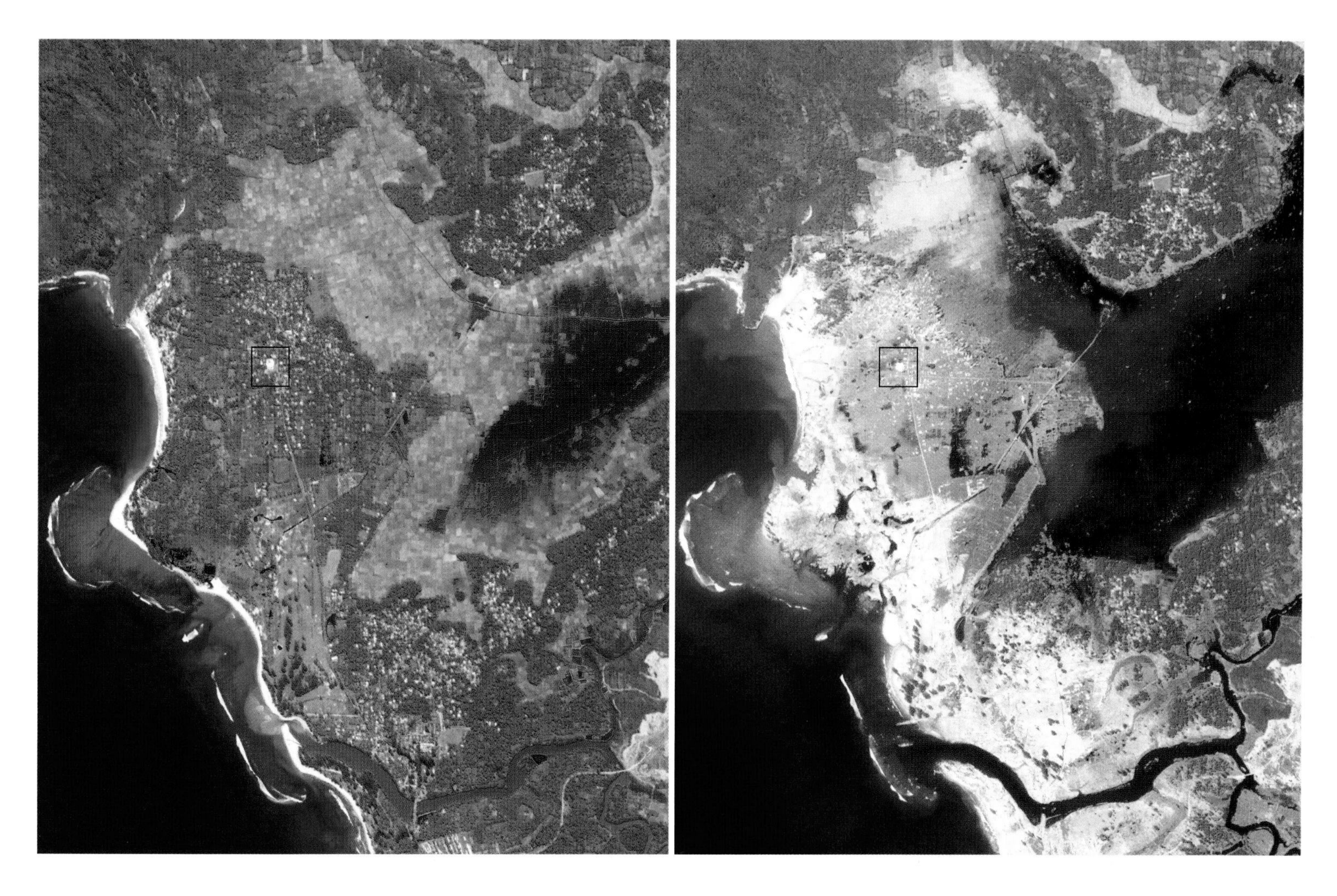

NEXT PAGE:
Banda Aceh, Sumatra, Indonesia in January 2003 (left) before the tsunami of 26 December 2004 and then soon afterwards on 29 December 2004 (right). Much of the coast is under water, vegetation has been stripped off and many buildings have been destroyed.

BELOW:
On 26 December 2004 an earthquake off the coast of Sumatra, Indonesia triggered the tsunami which then rapidly spread across the whole width of the Indian Ocean, reaching the coast of Somalia, over 3,100 miles (5,000km) away, seven hours after the earthquake.

Tsunami travel times

Time taken for tsunami to travel across Indian Ocean (hours)

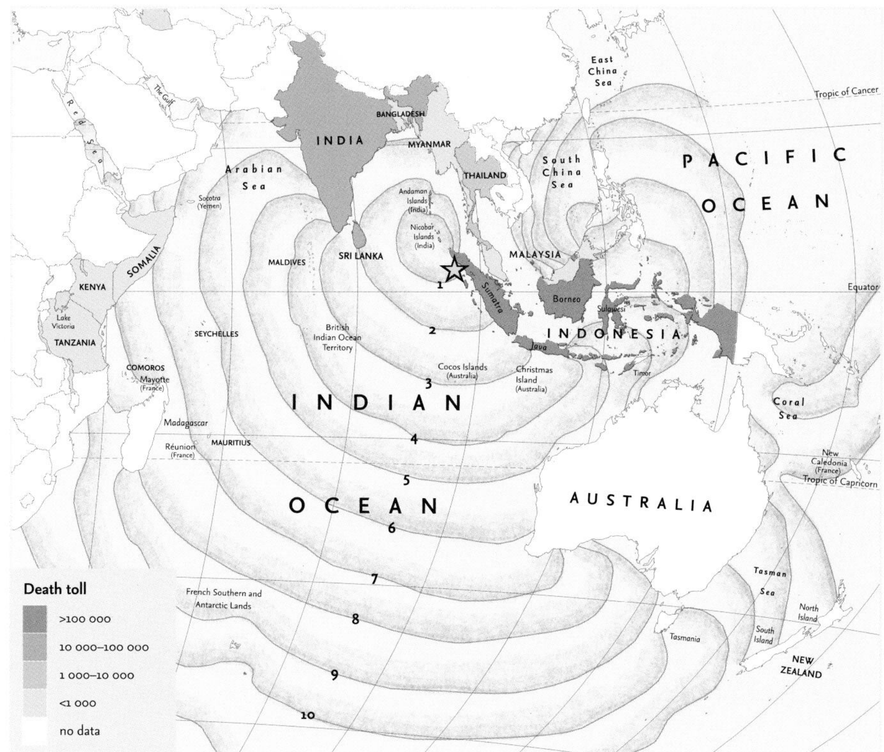

Indian Ocean tsunami death tolls

Estimated death tolls by country (March 2006)

Country	Death toll
Indonesia	165 708
Sri Lanka	35 399
India	16 389
Thailand	8 345
Somalia	298
Maldives	102
Malaysia	80
Myanmar	71
Tanzania	10
Seychelles	3
Bangladesh	2
Kenya	1
Total	226 408

Storm surge

Storm surge is simply water which is pushed toward the shore by the force of the winds swirling around a storm. This advancing surge combines with the normal tides to create a storm tide, which in the case of intense storms such as hurricanes can increase the average water level by 15 ft (4.5 m) or more.

In addition, wind-driven waves are superimposed on the storm tide. This rise in water level can cause severe flooding in coastal areas, particularly when the storm tide coincides with the normal high tides. Because much of the United States' densely populated Atlantic and Gulf Coast coastlines lie less than 10 ft (3 m) above mean sea level, the danger from storm tides is tremendous.

The level of surge in a particular area is also determined by the slope of the continental shelf. A shallow slope off the coast will allow a greater surge to inundate coastal communities. Communities with a steeper continental shelf will not see as much surge inundation, although large breaking waves can still present major problems. Storm tides, waves, and currents in confined harbors severely damage ships, marinas, and pleasure boats.

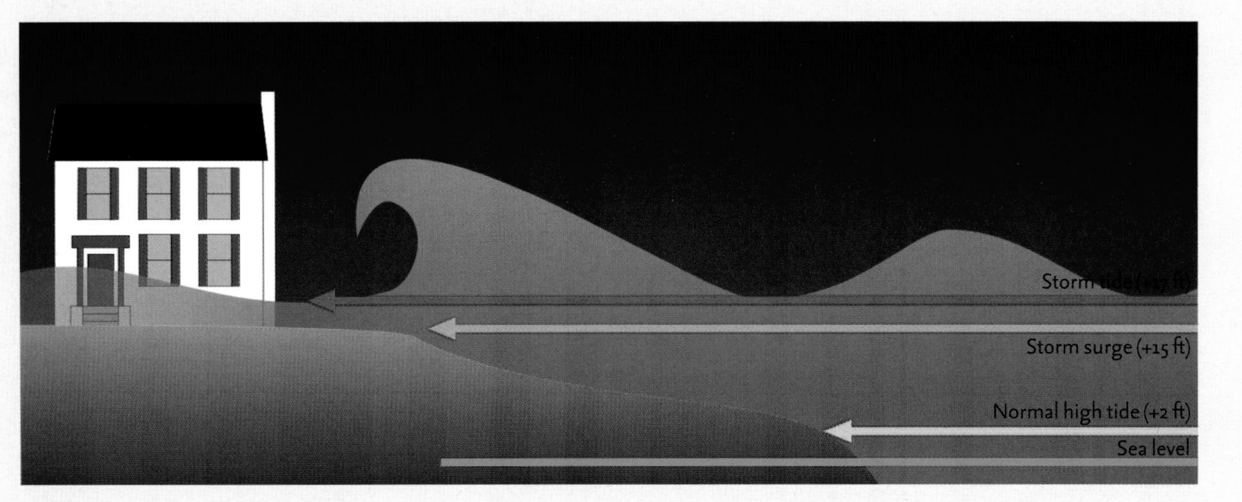

ABOVE:
Storm surge damage to beach front homes on Dauphin Island, Alabama, U.S.A..

LEFT:
A storm surge combined with a normal high tide results in a storm tide, increasing the average sea level by several feet and causing severe flooding in coastal areas.

Rogue waves

There are many sailors' tales of "rogue waves," "freak waves," "monster waves," "three sisters," and other "killer waves." Rogue waves are loosely defined as waves which are more than two times the significant wave height. Characteristics of rogue waves include having heights greater than twice the size of surrounding waves, coming from unexpected directions, and unpredictability. Rogue waves tend to appear in two types: Type One: Non dispersive, steep fronted, uniformly long crested, described as a "wall of water," and Type Two: Dispersive, elevating above the existing sea, sharply breaking, short lived. These typically occur with convective high winds.

Waves of over 98 ft (30 m) in height have been reported and are described as steep-sided "walls of water" with unusually deep trough. The USS Ramapo reported one such wave with a height of 112 ft (34 m) in the Pacific in 1933. One such wave in the north Atlantic Ocean nearly capsized RMS Queen Mary during World War II while it was carrying over 15,000 troops.

These enormous waves generally form because swells, while traveling across the ocean, do so at different speeds and directions. As these swells pass through one another their crests, troughs, and lengths randomly coincide and reinforce each other, combining to form unusually large waves which tower then disappear. If the swells are traveling in the same direction, these mountainous waves may last for several minutes before subsiding.

It is very seldom that huge waves over 65 ft (20 m) are developed; and, fortunately, sailors today normally do not encounter them as ships try to avoid such conditions by altering course before the storm.

ABOVE:
The Thames flood barrier, London, U.K. with the gates closed. The gates are closed when there is a high risk of tidal flooding, or for maintenance. Closing the barrier seals off part of the upper Thames with a continuous steel wall spanning 1,700ft (520m) across the river.

LEFT:
Large waves pounding one of NOAA's ships during an expedition.

Rip currents

Rip currents are channeled currents flowing away from shore. They typically extend from the shoreline, through the surf zone, and past the line of breakers. Rip currents can occur at any beach with breaking waves.

Rip currents typically form at low spots or breaks in sandbars, and also near structures such as groins, jetties and piers. Rip currents can vary in width from a few feet to over 300 feet (100 meters). Depending on conditions, the seaward pull of rip currents can extend from just beyond the line of breakers to hundreds of yards offshore.

They form as incoming waves break over a sandbar and raise the water level between the bar and shore. This causes currents parallel to the shore as the waves push more and more water in between the sandbar and the shore. These currents converge and rush back to sea as a rip current through breaks or channels through the sandbar. Once the rip current passes through the narrow gap it begins to weaken considerably.

Rip currents are a leading hazard for beachgoers: the United States Lifesaving Association estimates that the annual deaths due to rip currents on U.S. beaches exceeds 100. They also account for over 80 per cent of rescues performed by lifeguards. They are particularly dangerous for weak or non-swimmers. Rip current speeds are typically 1–2 feet per second (0.3–0.6 meters per second), but stronger ones can attain velocities as high as 8 feet per second (2.4 meters per second), faster than an Olympic swimmer can sprint!

A common myth concerning rip currents is that they are associated with an undertow. This is wrong as the rip current is a horizontal current moving away from the shore. If caught in a rip current, do not panic. Swim parallel to the shore, not against the current, until escaping the grip of the rip current, and then angle in to the shore.

ABOVE:
Flags are being used at Miami Beach, Florida, U.S.A. to warn people of moderate surf/currents (yellow flag) and dangerous marine life (purple flag). Dangerous marine wildlife usually means jellyfish and other poisonous animals.

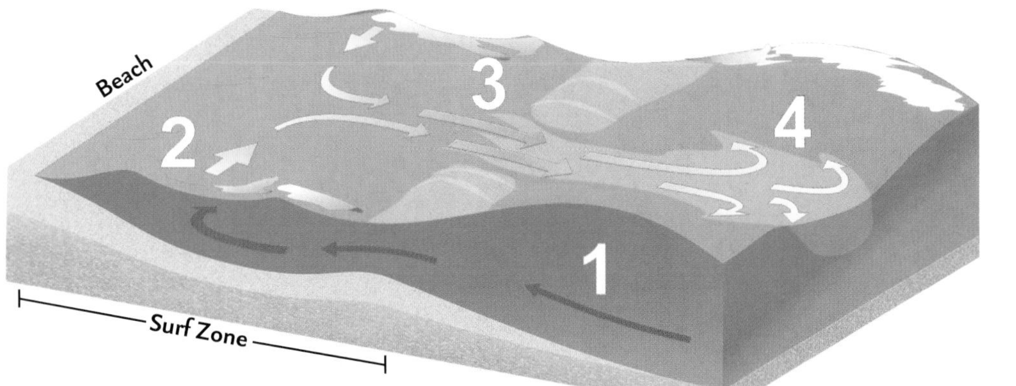

1. Incoming waves move water past an underwater sandbar.
2. Water collects between sandbar and shore.
3. Rip current forms as weight of water creates gap in sandbar with water rushing seaward.
4. As the current of water passes sandbar it begins to spread out, decrease in velocity and strength, and eventually dissipates.

Waterspouts

Waterspouts, which are similar to tornadoes over water, are generally broken into two categories: fair weather waterspouts and tornadic waterspouts.

Tornadic waterspouts are simply tornadoes which form over water, or move from land to water. They have the same characteristics as land tornadoes. They are associated with severe thunderstorms, and are often accompanied by high winds and seas, large hail stones, and frequent dangerous lightning.

Fair weather waterspouts are usually a less dangerous phenomenon, but common over tropical waters. The term "fair weather" comes from the fact that this type of waterspout forms during fair and relatively calm weather, often during the early to mid-morning and sometimes during the late afternoon. Fair weather waterspouts usually form along dark flat bases of a line of developing cumulus clouds. This type of waterspout is generally not associated with thunderstorms, whereas tornadic waterspouts develop in severe thunderstorms. Tornadic waterspouts develop downward in a thunderstorm while a fair weather waterspout begins to develop on the surface of the water and works its way upward. By the time the funnel is visible, a fair weather waterspout is near maturity.

Fair weather waterspouts form in light wind conditions so they normally move little. If a waterspout moves onshore, the National Weather Service issues a tornado warning as some of them can cause significant damage and injuries to people. Typically, fair weather waterspouts dissipate rapidly when they make landfall, and rarely penetrate far inland.

The best way to avoid a waterspout is to move at a 90° angle to its apparent movement. Never move closer to investigate a waterspout. Some can be just as dangerous as tornadoes and even fair weather waterspouts can be dangerous to small craft.

LEFT:
A waterspout forms in the Adriatic Sea off the coast of Croatia. The air within the cloud has started to rotate and is emerging from it as a funnel (top). Gradually this funnel extends downwards until it reaches the surface of the sea and the waterspout is formed (bottom). The wind speed on the edge of this waterspout reached 66 mi/hr (170km/hr).

Biological hazards in the sea

There are a variety of natural biological hazards in the sea ranging from huge creatures which can attack and eat people to microscopic creatures that contain harmful toxins. Many marine organisms will attack humans or defend themselves vigorously if threatened or disturbed; some inject poison if touched, bitten by, or stepped on (with symptoms ranging from mildly irritating to fatal through a variety of mechanisms including spines, harpoon-like darts, fangs, and nematocysts); others cause illnesses ranging from mild intestinal distress through paralysis to death if ingested.

Many ocean carnivores have been known to attack humans. Dangerous carnivores include the great white shark, tiger shark, oceanic white-tip shark, and bull shark, various cephalopods, notably the Humboldt squid, and the saltwater crocodile. Many other sea creatures have been known to injure humans under certain conditions including various marine mammals, moray eels, barracuda, and large schooling fish such as bluefish and mackerel.

Poisonous sea creatures include: various cnidarians such as box jellyfish, Portuguese men-of-war, fire corals, and various anemones; asteroidea including some types of sea urchins and the crown-of-thorns starfish; various mollusks including the blue-ring octopus, and cone shells, in particular Conus geographicus; some types of sponges; sea snakes; fire worms; members of the scorpion-fish family including stonefish, scorpion-fish, and lion-fish; and sting-rays. Australian waters are home to two of the most poisonous creatures on earth, the box jellyfish and the blue-ringed octopus. If a person is injected with sufficient poison from either of these creatures, death can occur within minutes.

Fortunately, most marine animal incidents involving humans do not result in fatal consequences. However, there have been infamous incidents involving maritime disasters that have resulted in multiple shark-attack fatalities. However, it has been estimated that there are at least 150 sea-snake envenomation fatalities per year as compared to 10–15 shark attack fatalities per year.

TOP ROW:
Scorpion-fish, sea urchin, barracuda, moray eel.
MIDDLE ROW:
White-tip shark, lionfish, stingray.
BOTTOM ROW:
Crown of thorns starfish, sea snake, purple striped jellyfish.

The Poles

The Poles
Introduction

Polar bears in the Arctic and penguins in Antarctica – that is one of the most popular distinctions the public first make about the polar regions. In fact, as they relate to the geophysical nature of the polar oceans and their interaction with the atmosphere and the planet's climate, the differences between the Arctic and the Antarctic regions are far more reaching. The Arctic Ocean includes the North Pole and is surrounded by Europe, Asia, and North America. It is the smallest of the earth's oceans with a total area of approximately 5.4 million square miles (14 million sq km). There is restricted flow of sea water and ice between the Arctic Ocean and the Bering Sea through the relatively narrow and shallow Bering Strait. Additional outflow occurs through the Nares Strait between Greenland and Ellesmere Island and through the Canadian archipelago into Baffin Bay. The most important exchange of Arctic waters occurs with the Atlantic Ocean between Greenland and Norway with the outflow of water and sea ice through the Fram Strait east of Greenland and the inflow of warmer water north of Norway. Significant amounts of fresh water are also received by the Arctic Ocean from the Mackenzie river in Canada and the Ob, Yenisey, and Lena rivers in Siberia. The Arctic Ocean maintains density stratification (a slight increase of salinity with depth). This results from the fresh water input and the process of sea water freezing and melting.

The Southern Ocean is defined as the southern portion of the Atlantic, Pacific, and Indian Oceans beyond latitude 60°S. With an area of 7.7 million square miles (20 million sq km), it surrounds the 5.4 million square miles (14 million sq km) of the Antarctic continent, which includes the South Pole. Because of the geographical differences between the polar oceans, and the fact that, as opposed to the Arctic region, there are no indigenous people in Antarctica, ocean exploration and our present approach to research, resource exploitation, and environmental management in each region have taken different paths and faced unique challenges. The major Southern Ocean circulation is defined by the west to east flow of the Antarctic Circumpolar Current (ACC), the world's largest ocean current. The Southern Ocean geophysical extent is also delimited by the location of the Antarctic Convergence region formed where the ACC cold waters meet with warmer waters to the north. The Southern Ocean is much deeper than the Arctic, with a depth of 13,100–16,400 ft (4,000–5,000 m) over most of its extent with generally narrow and deep Antarctic continental shelves, in contrast to the wide and relatively shallow Siberian continental shelf in the Arctic.

PREVIOUS PAGE:
Bering Strait, April 22 (left) and May 22, 2002 (right). Seasonal thawing of the Bering Strait between Siberia on the left and Alaska on the right hand side of each image.

LEFT:
Elephant Island at sunrise in the Southern Ocean.

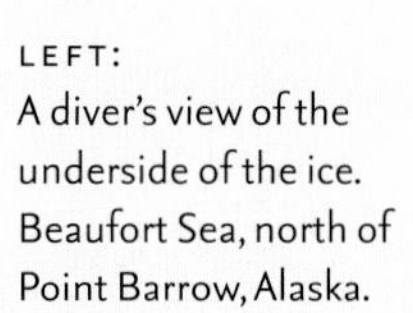

LEFT:
A diver's view of the underside of the ice. Beaufort Sea, north of Point Barrow, Alaska.

LEFT:
An Arctic cod rests in an ice-covered space.

Arctic Ocean human impacts

There are eight nations which lie within the Arctic. These nations, U.S., Russia, Canada, Greenland (Denmark), Iceland, Norway, Sweden and Finland are populated by a mix of people ranging from indigenous to more recent inhabitants. The indigenous population is heavily dependent on the natural resources in the region, and some conflicts have arisen with governments eager to exploit these rich resources. There have also been moves towards greater autonomy for such indigenous groups. Most notably, the creation of the new Canadian territory of Nunavut, in 1999, recognized Inuit land claims and harvesting rights.

The United States, Russia, Canada, Norway, and Denmark are at the center of a mounting issue over who owns the seafloor territory of the Arctic Ocean, potentially rich with oil and gas. Global warming is another issue which is highlighted in the Arctic. Sea ice cover is markedly retreating to the extent that it is likely that new shipping routes could be opened up allowing new areas to be exploited. This will put more pressure on the delicate polar environment. Fresh water changes from melting glaciers, and increased river input could also cause changes in the global thermohaline circulation, potentially impacting the entire northern hemisphere.

Due to its geographical and climatic characteristics, the Arctic is at risk from many pollutants from all over the planet and pollution is becoming an increasingly serious problem. Contaminants such as radioactive material, emissions from industry, organic toxins, and heavy metals are transported to the Arctic through atmospheric, oceanic, and land-based routes. The harsh environment has led to flora and fauna being uniquely and finely adapted to their environment, which means they are particularly vulnerable to such contamination. Animals high up the food chain such as polar bears, seals, and whales often have high concentrations of these pollutants in their tissue.

The Arctic and adjacent seas have some of the richest fishing grounds in the world. However, many species have been seriously over fished and are now under threat. Increasing numbers of immature fish are being caught, further impacting fish stocks. Steps are being taken by several countries to sustain the fishing industry. One example of this is in the Barents Sea where improved cooperation between the states bordering the sea has led to the establishment of no-fishing zones and quota systems to allow fish numbers to recover.

Increasing numbers of tourists are visiting the Arctic. As well as the usual environmental risks presented by such tourism, these visits also help to increase awareness of the environmental issues of the region.

ABOVE:
U.S. Coast Guard Cutter Healy (an icebreaker) crew and science team during NOAA-sponsored 2005 Hidden Ocean Expedition to the Arctic.

LEFT:
Los Angeles-class fast attack submarine USS Alexandria is partially submerged after surfacing through two feet of ice during a U.S. Navy and Royal Navy exercise conducted on and under a drifting ice floe about 180 nautical miles (333 km) off the north coast of Alaska, U.S.A.

RIGHT:
Scientists extracting an ice core and taking measurements on a core already brought to the surface. These ice cores provide valuable information to help with future climate predictions.

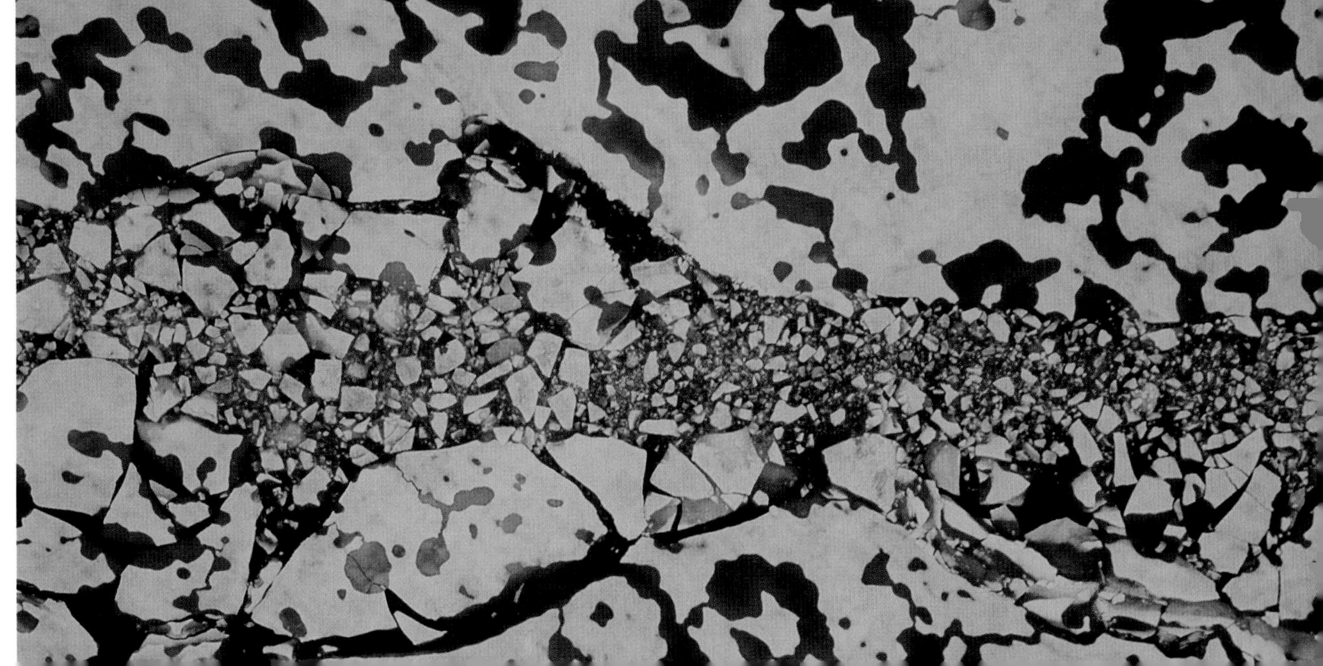

RIGHT:
Ponded sea ice which was traversed by U.S. Coast Guard Cutter Healy during its 2005 Trans-Arctic Transect as seen from a polar-orbiting

Arctic Ocean sea ice extent

A significant decrease in Arctic sea ice extent, thickness, age of multi-year ice (MYI), and the ratio of MYI to first year ice (FYI) have been observed in the last few decades with an apparent acceleration in the last five to six years. In general, summer melt appears to be occurring earlier in the season, with freeze-up occurring later in the fall. Even more dramatic has been the massive loss of MYI being experienced through the Fram Strait, particularly in the last three years. The high rate of MYI or perennial ice loss has set the stage for record surface areas of the Arctic Ocean to be ice-free during the summer months – a process happening much faster than predicted by climate models. These sea ice changes and the overall warming of the Arctic Ocean are already having significant effects in the regional weather and climate, marine ecosystems, and a new found maritime accessibility to this region of the world.

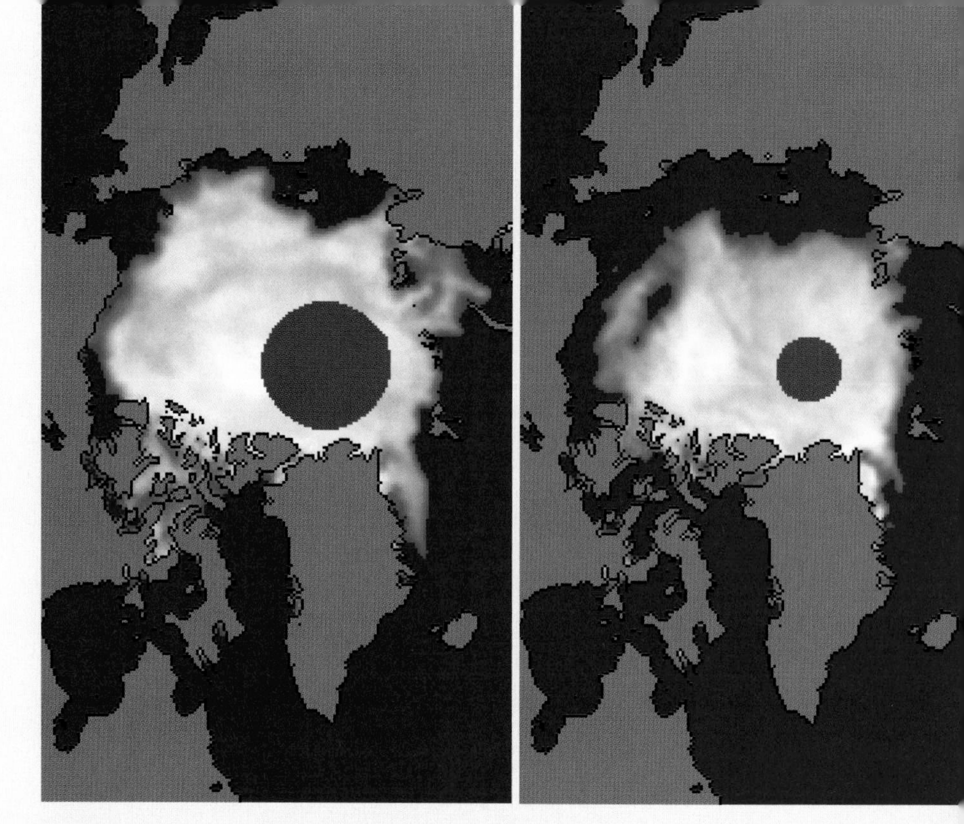

ABOVE:
Arctic sea ice concentration during summer minimum extent conditions in September 1980 (left) and September 2006 (right). Lightest areas show the greatest concentration of ice, dark blue areas show the least.

BELOW:
Aerial view from aircraft of melting sea ice floes.

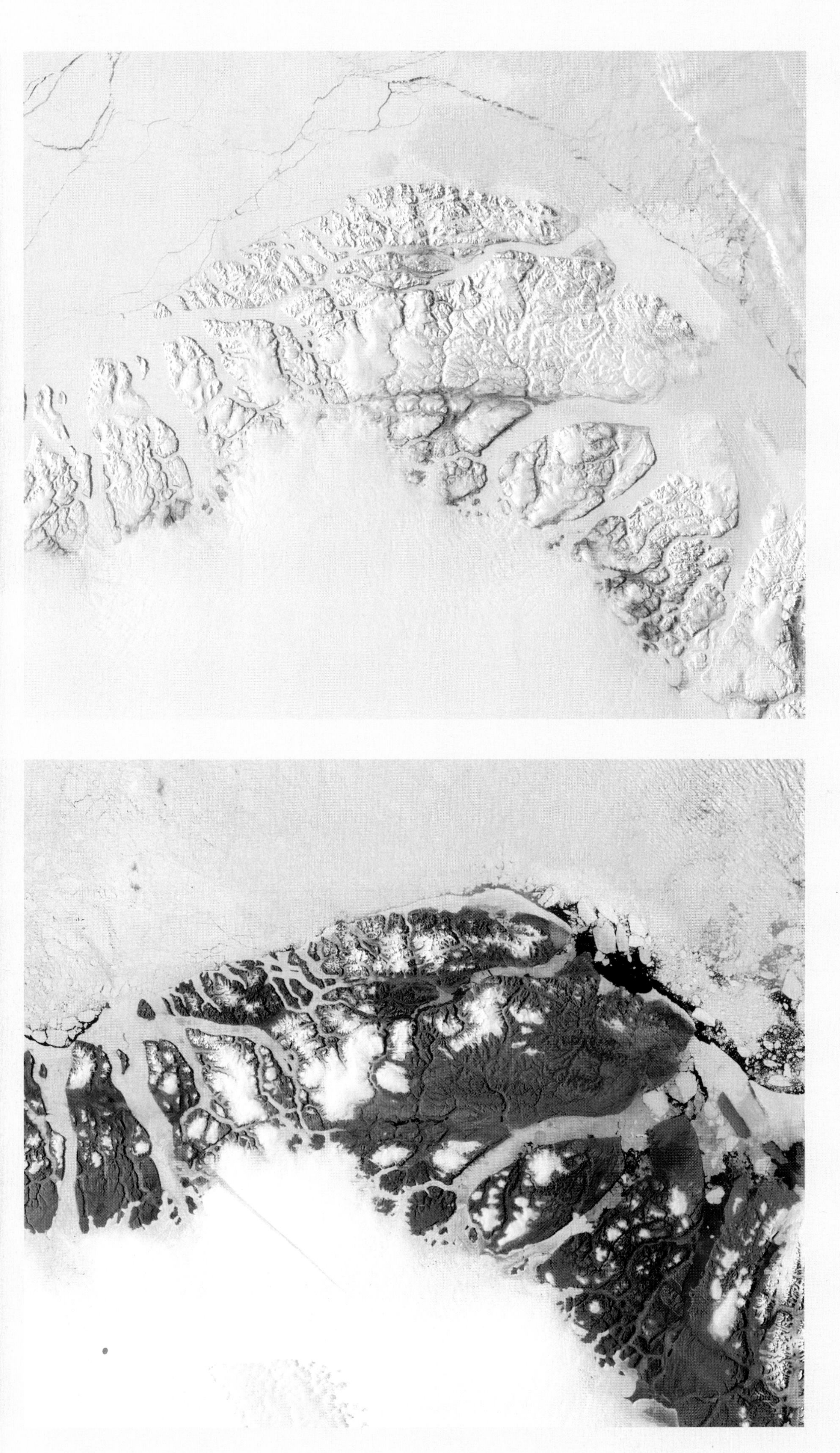

LEFT:
Springtime in northeastern Greenland as snow and ice retreat from the rugged coastline.

LEFT:
The R/V Roger Revelle, 273 ft (83 m) in length, seen as it nears an iceberg in the Southern Ocean near 65°S.

ABOVE:
Antarctic toothfish, which are marketed as Chilean sea bass, are currently the most economically important resource harvested in the Southern Ocean.

Antarctica human impacts

Much of Antarctica and the Southern Ocean remains pristine. Nevertheless, humans have directly influenced parts of both the terrestrial and marine environment through exploration, scientific endeavors, tourism, and most significantly, commercial exploitation of marine living organisms. Although the Southern Ocean has historically teemed with life, animals which inhabit these waters are among the earth's most vulnerable.

Harvesting of Antarctic life began in the 1820s with fur seals, then with whales in the early 1900s. Within a few decades, these groups were driven to near extinction. More recently (since the 1970s) humans have targeted krill, crabs, squid, and fish. During the first decade of this period, several stocks of fish were exploited in the absence of regulation, and were driven to such low levels that their harvesting was no longer commercially viable, and many fisheries were closed.

Since the early 1980s, there has been a concerted international effort to manage Antarctic living marine resources more wisely, using an approach with an objective of maintaining ecological relationships between harvested, dependent, and related species. The commercial fishing industry in the Southern Ocean is now relatively well regulated, and targets primarily Antarctic krill and Patagonian/Antarctic toothfish, widely marketed as Chilean seabass.

Presently, the population of Antarctic krill, from which a variety of products can be derived, has been little influenced by human harvest. This is due primarily to their very large biomass and high productivity, coupled with the fact that they are difficult to process and have a relatively limited market. In contrast, toothfish have experienced a rapid increase in exploitation during the last decade, and are a highly valuable seafood product. Furthermore, in some parts of the Southern Ocean, there is illegal, unregulated, and unreported fishing. Nevertheless, using sound scientific principles and an ecosystem approach, there are areas where toothfish stocks are well managed, and can be harvested on a long term, sustainable basis.

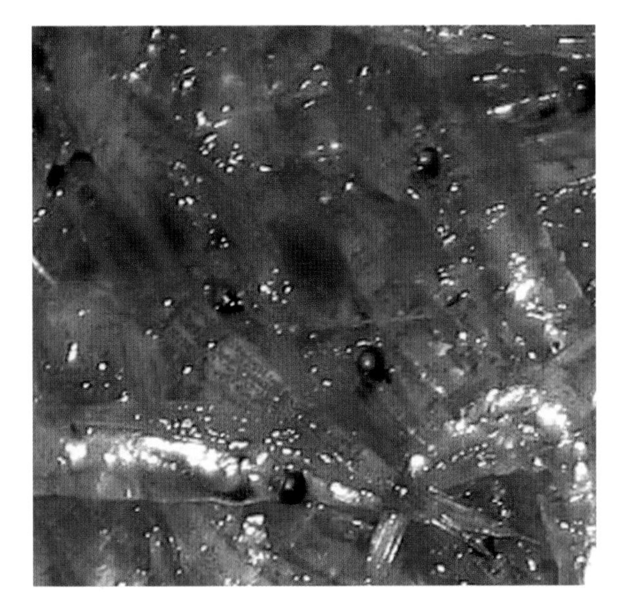

ABOVE AND TOP:
Antarctic krill are a keystone species in the Antarctic ecosystem. To date, their overall numbers have been little influenced from exploitation. This may change rapidly as the human population grows around the world and new markets for krill develop.

TOP RIGHT:
A cargo ship in icy waters.

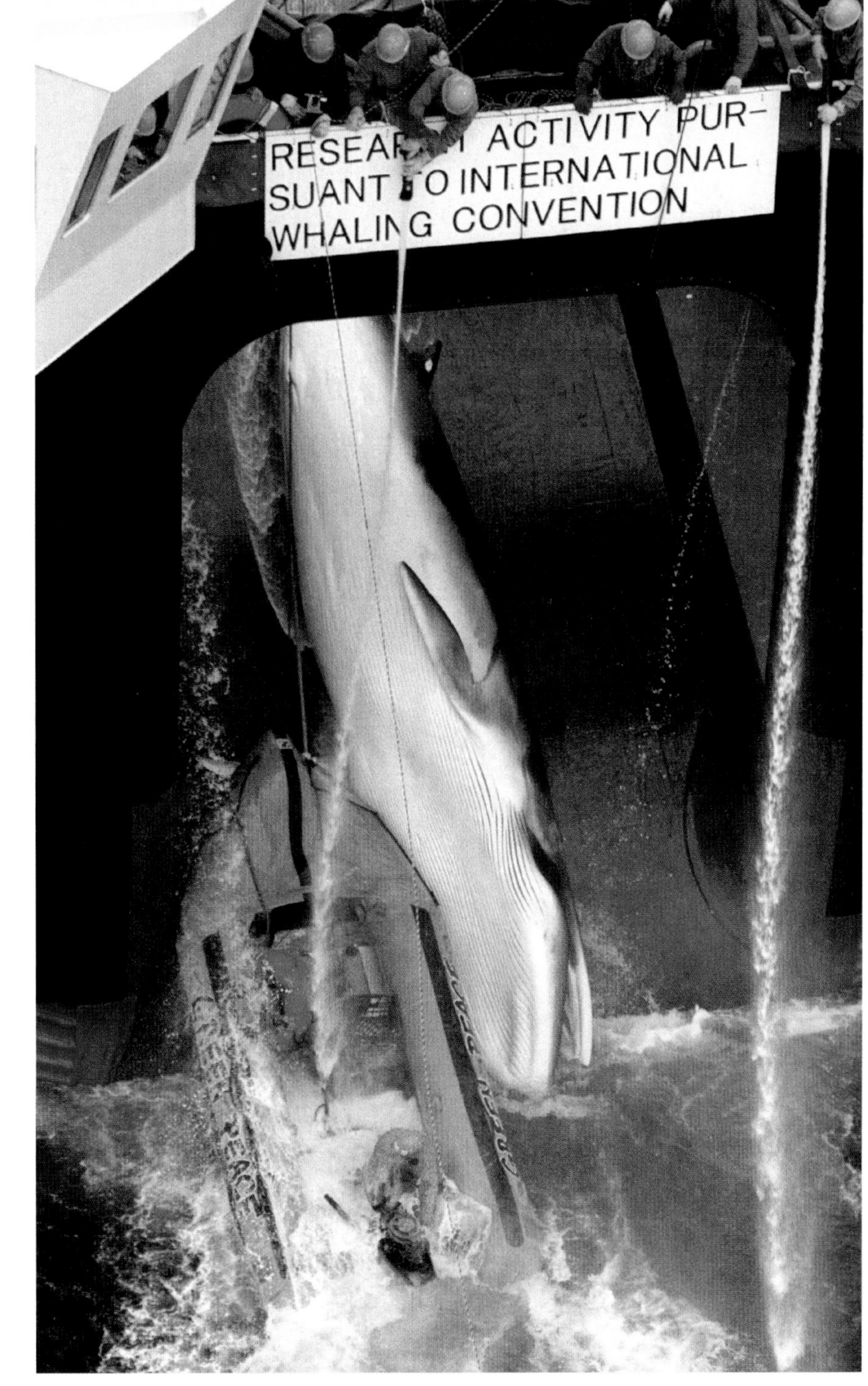

RIGHT:
At least five of the thirteen great whales are listed as endangered due to whaling. The International Whaling Commission (IWC) was set up to monitor the situation and protect whale numbers. This image shows Greenpeace demonstrators disrupting a whale catch in the Southern Ocean Whale Sanctuary.

Antarctica sea ice extent

Most of the sea ice around Antarctica is seasonal with a minimum ice pack extent of less than 1.2 million square miles (3 million sq km) in March, growing to approximately 7.5 million square miles (19 million sq km) in September. The marine ecosystems in each polar region are highly influenced not just by the presence of sea ice but by the very different conditions in terms of ice type, extent, and seasonality. The ocean around Antarctica is also characterized by the presence of a large number of icebergs produced by the break up of Antarctic ice shelves and glaciers. These can include huge icebergs several miles in length.
The Southern Ocean is protected by international Antarctic agreements which aim to regulate or prohibit fisheries, whaling, environmental pollution, mineral resource exploitation, research, and other activities by sovereign states.

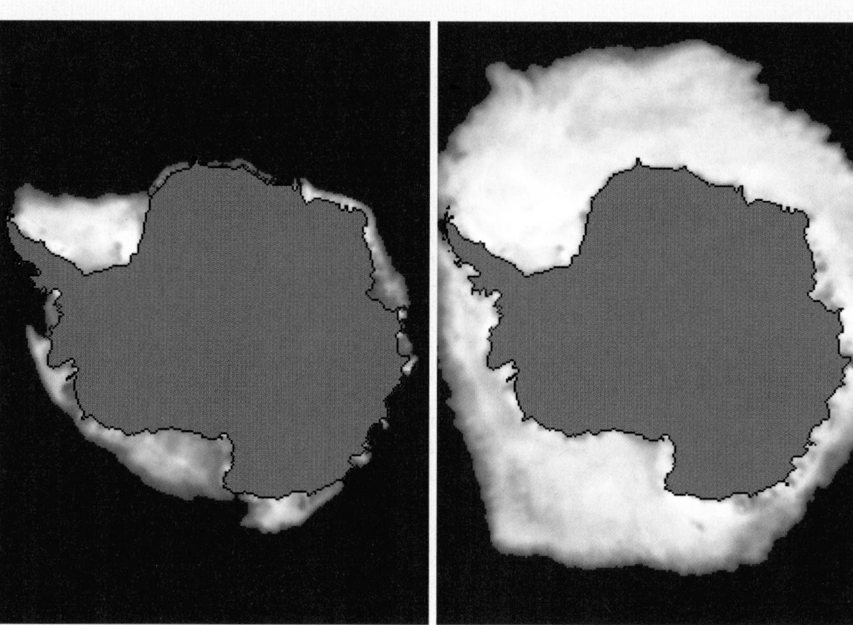

ABOVE:
Antarctic sea ice concentration during March 2006 minimum extent conditions (left) and September 2006 maximum extent conditions (right). Lightest areas show the greatest concentration of ice, dark blue areas show the least.

TOP:
Sunset over sea ice at the marginal ice zone.

LEFT:
The Ross Sea, Antarctica.

RIGHT:
Open water wake in the polar sea ice.

RIGHT:
Pancake ice adrift on the Ross Sea.

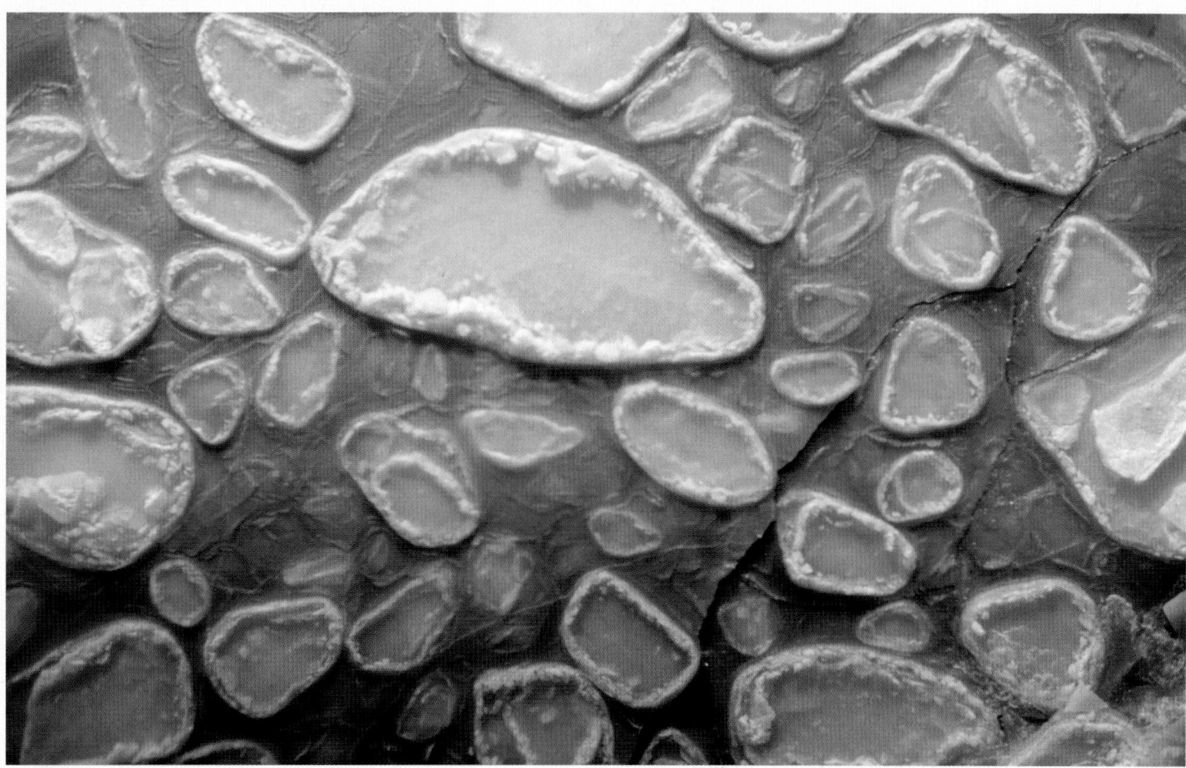

RIGHT:
Edge of the Ross Ice Shelf.

Polar life

ABOVE: Clione (sea butterfly), polar bear cub and mother (Arctic), killer whales, humpback whale, spotted seal pup, deep red medusa, and emperor penguins and chicks (Antarctica).

Arctic Ocean

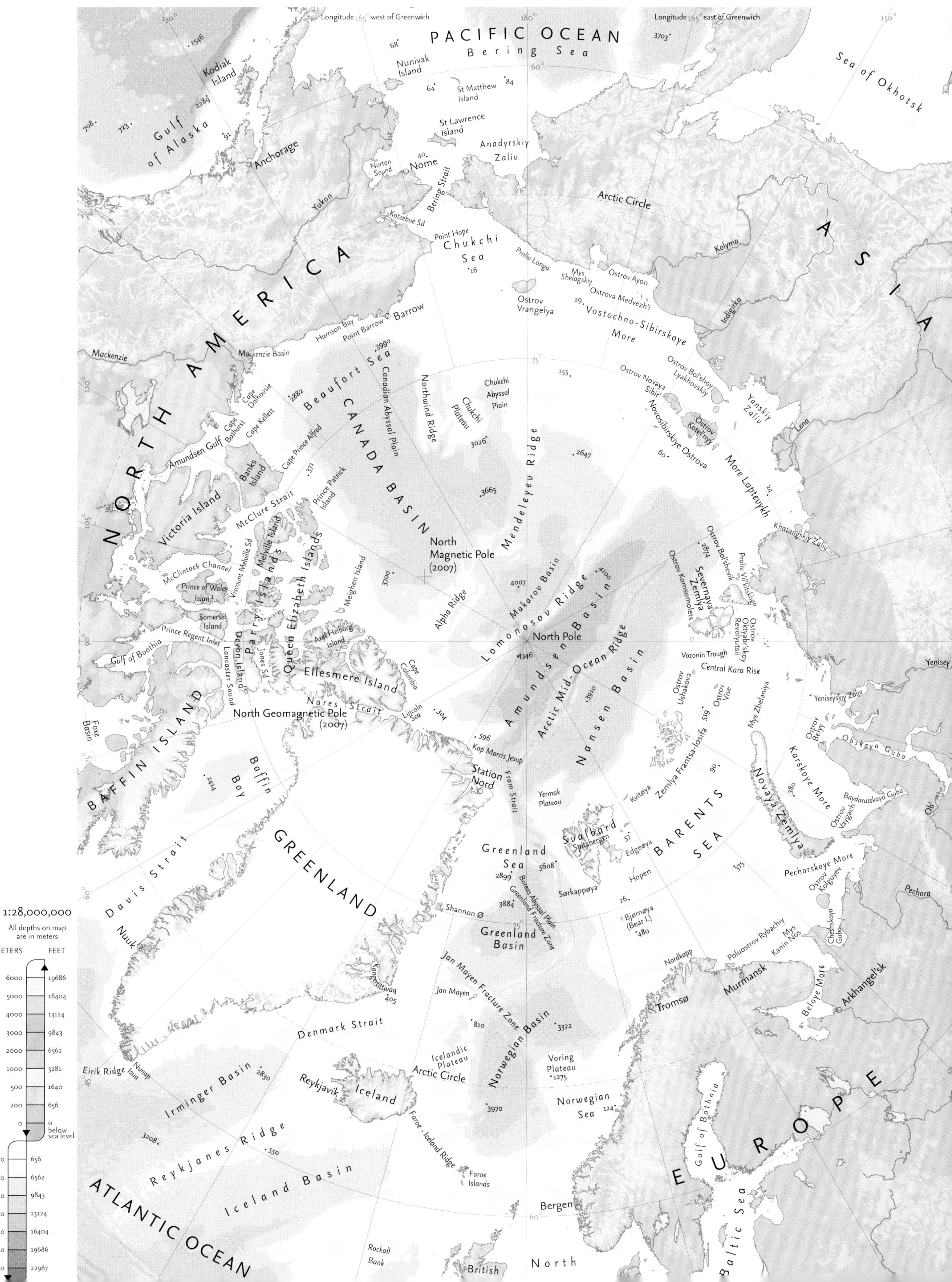

Southern Ocean

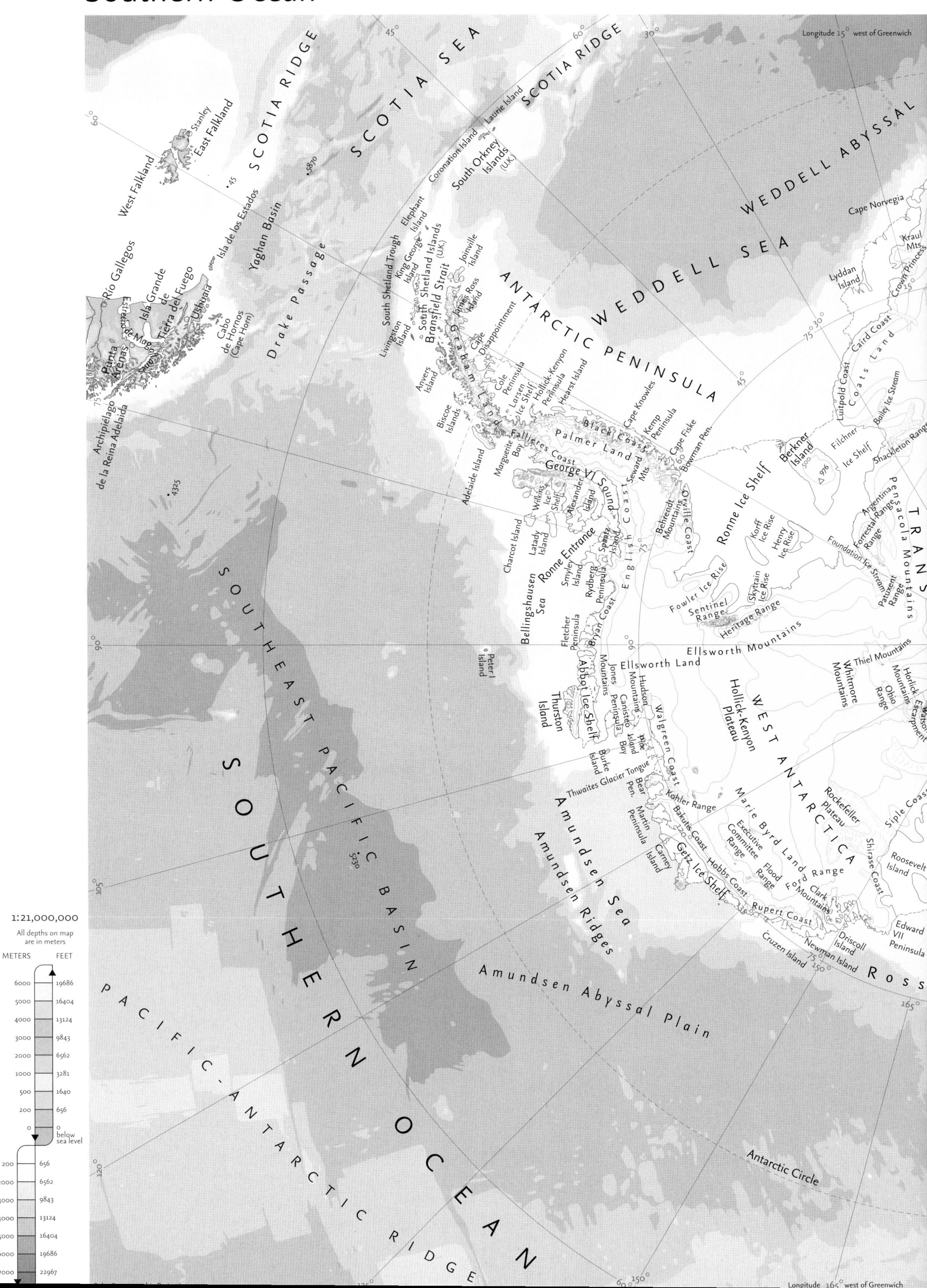

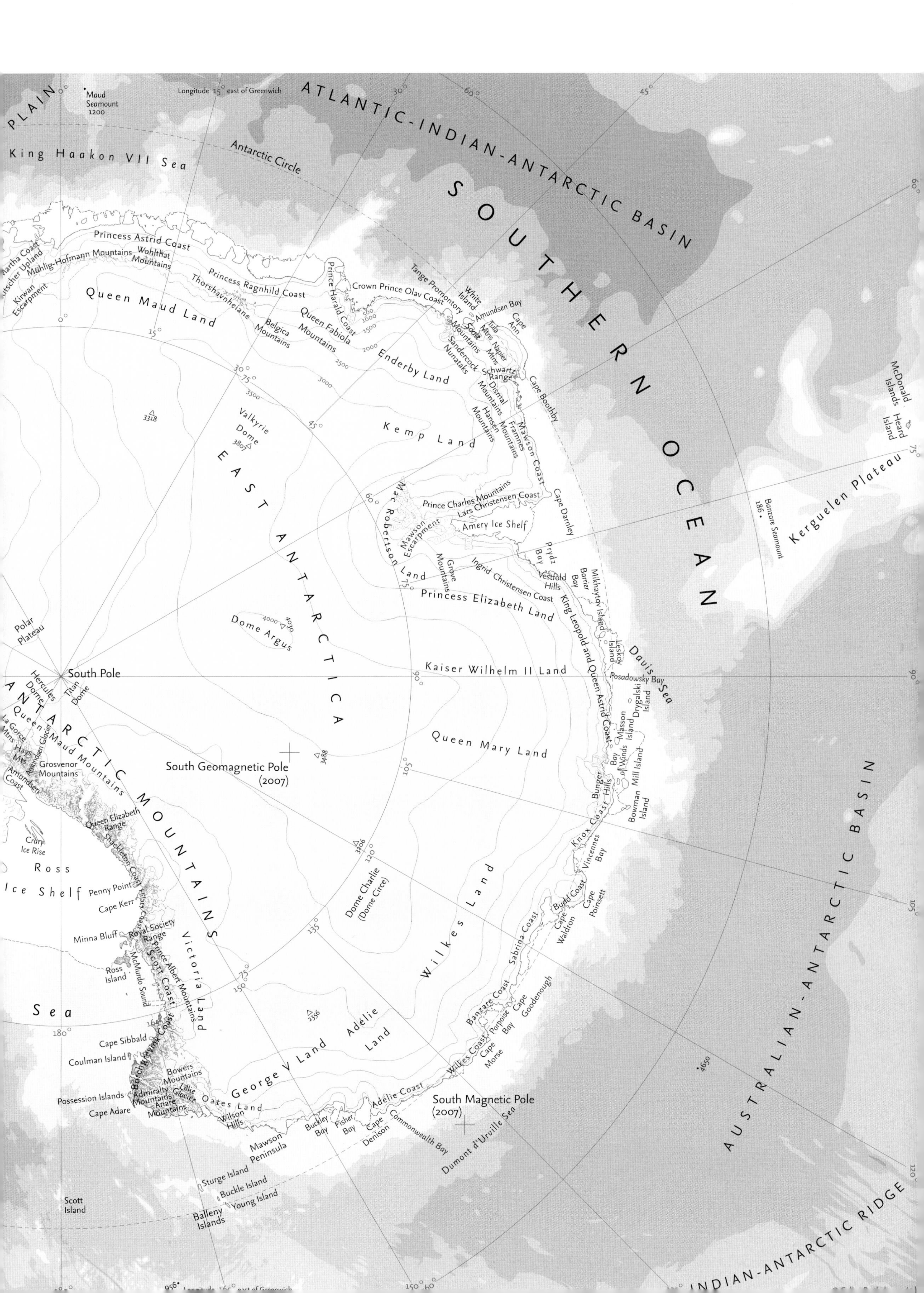

SOUTHERN OCEAN
ATLANTIC-INDIAN-ANTARCTIC BASIN
AUSTRALIAN-ANTARCTIC BASIN
INDIAN-ANTARCTIC RIDGE
EAST ANTARCTICA
TRANSANTARCTIC MOUNTAINS
King Haakon VII Sea
Antarctic Circle
Maud Seamount 1200
Longitude 15° east of Greenwich
Princess Astrid Coast
Princess Ragnhild Coast
Queen Maud Land
Enderby Land
Kemp Land
Mac Robertson Land
Princess Elizabeth Land
Kaiser Wilhelm II Land
Queen Mary Land
Wilkes Land
Adélie Land
George V Land
Oates Land
Victoria Land
Kerguelen Plateau
Banzare Seamount 186
McDonald Islands
Heard Island
Davis Sea
Dumont d'Urville Sea
Ross Ice Shelf
Ross Sea
South Pole
Polar Plateau
Dome Argus
Valkyrie Dome
Dome Charlie (Dome Circe)
South Geomagnetic Pole (2007)
South Magnetic Pole (2007)
Amery Ice Shelf
Prydz Bay
Mawson Coast
Cape Darnley
Scott Island
Balleny Islands
Sturge Island
Buckle Island
Young Island

Vital ecosystems

Vital ecosystems
Introduction

While the world's oceans cover more than 70 per cent of the earth's surface, the majority of species are found in bands along the margins of the continents. In these areas, light penetrates down through the water and provides the fuel which drives the basic productivity of these ecosystems. All these nearshore and coastal ecosystems begin with plants, but the form of these plants takes varies widely from giant kelp forests to lush beds of seagrass to single celled algae imbedded in the tissues of coral. The plants, then, provide food, nesting and spawning sites, refuge from predators, and homes to animals of all ages and sizes. Individual ecosystems are interconnected with those adjacent, through the exchange of water, organisms, and nutrients. This interconnection ranges in scale from linkages between a seagrass bed and adjacent open water areas, to the large-scale flow of nutrients, sediments, and organisms through large marine ecosystems such as the Gulf of Mexico. These intertwined and highly productive ecosystems support much of what we treasure about the oceans, from food sources which support cultures worldwide to intangibles which improve the quality of our lives.

This chapter will explore some of the critical ecosystems found along the oceans' margins, introducing the reader to what makes these areas important and discussing some of the challenges in protecting or restoring the habitats while preserving our ability to live and work near and in them.

RIGHT
Ord River Delta in Western Australia, with mangroves lining the water channels. Salt water from the incoming sea mixes with the fresh water of the river and makes it brackish. Mangrove vegetation is adapted to survive the high salinity of the water, and their roots allow gaseous exchange to counteract the poorly aerated soil.

LEFT
Sea anemone attach themselves to rocks and provide a habitat for a wide variety of species such as clownfishes, anemone shrimps, and anemone crabs.

PREVIOUS PAGE
Aerial view of islands amongst the sand bars of the Bahama Bank, an extremely shallow region of the Caribbean Sea. A large tidal estuary-like channel can be seen between the two main islands, extending into a large, shallow sand bar.

Large marine ecosystems

Large marine ecosystems (LMEs) are regions of ocean space of 77,220 sq miles (200,000 sq km) or greater. They encompass coastal areas from river basins and estuaries out seaward to the break or slope of the continental shelf (e.g. Northeast U.S. Continental Shelf LME), or out to the seaward extent of a well-defined current system (e.g. Guinea Current LME). Some, like the Black Sea LME, are semi-enclosed geographical areas. LMEs are defined by ecological criteria including bottom depth contours, currents and water mass structure, marine productivity, and food webs. As of 2007, sixty-four LMEs have been delineated (see map). The LMEs annually produce 95 per cent of the world's marine fish catch, and have varying degrees of productivity, influenced by natural causes and also by pollution and over exploitation of living resources by humans.

The LME approach to ecosystem-based management, developed by the National Oceanic and Atmospheric Administration, has identified three major driving forces which can cause change in the structure of an LME. These are overfishing, pollution, and climate change. The strategy for reducing coastal pollution, restoring damaged habitats, and recovering depleted fisheries depends on an integration of science and management at the LME scale. This strategy takes into account five areas: LME productivity, the physical and chemical characteristics, and persistent features of the LME over time; LME fish and fisheries, the current status of fisheries and changes in fish populations over time, determined through bottom trawls, hydro-acoustics, and satellite data; LME pollution and ecosystem health, the pressure from polluted sediments and excessive nutrients, for example; LME socioeconomics, the size and scope of activities by humans to exploit and manage ocean resources; and LME governance, the laws and regulations which govern ocean usage. In seventeen LME projects around the world funded by the Global Environment Facility, science brings 121 countries together to monitor the marine environment and identify the most pressing issues that need to be addressed and actions to be taken for sustaining LME resources.

LEFT:
Part of the southern portion of the Great Barrier Reef, adjacent to the central Queensland coast. The reef, delineated as a large marine ecosystem, extends for 1,240 miles (1,995 km) along the northeastern coast of Australia. It is not a single reef, but a vast maze of reefs, passages, and coral cays.

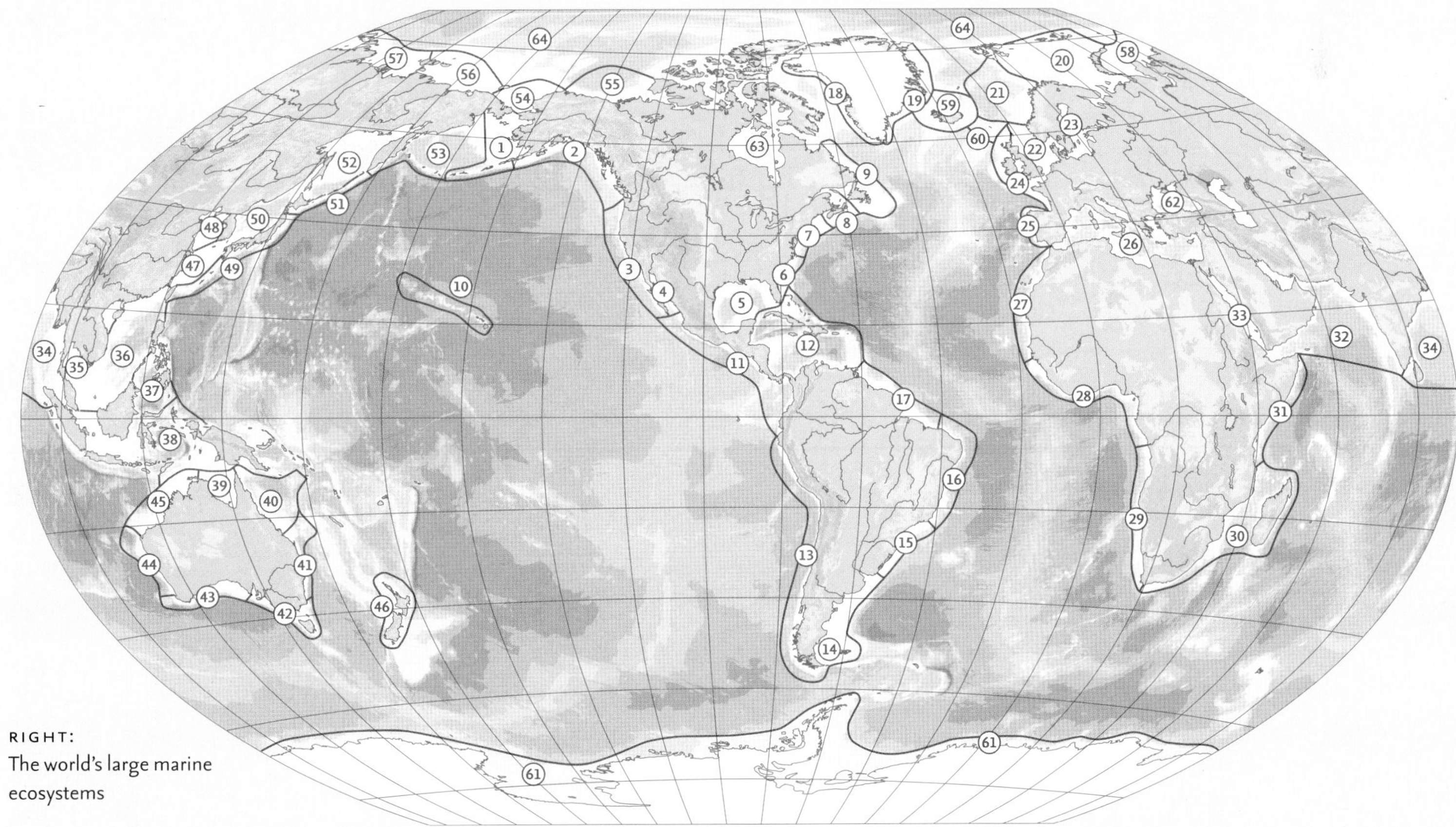

RIGHT:
The world's large marine ecosystems

1 East Bering Sea
2 Gulf of Alaska
3 California Current
4 Gulf of California
5 Gulf of Mexico
6 Southeast U.S. Continental Shelf
7 Northeast U.S. Continental Shelf
8 Scotian Shelf
9 Newfoundland-Labrador Shelf
10 Insular Pacific-Hawaiian
11 Pacific Central-American Coastal
12 Caribbean Sea
13 Humboldt Current
14 Patagonian Shelf
15 South Brazil Shelf
16 East Brazil Shelf
17 North Brazil Shelf
18 West Greenland Shelf
19 East Greenland Shelf
20 Barents Sea
21 Norwegian Shelf
22 North Sea
23 Baltic Sea
24 Celtic-Biscay Shelf
25 Iberian Coastal
26 Mediterranean Sea
27 Canary Current
28 Guinea Current
29 Benguela Current
30 Agulhas Current
31 Somali Coastal Current
32 Arabian Sea
33 Red Sea
34 Bay of Bengal
35 Gulf of Thailand
36 South China Sea
37 Sulu-Celebes Sea
38 Indonesian Sea
39 North Australian Shelf
40 Northeast Australian Shelf-Great Barrier Reef
41 East-Central Australian Shelf
42 Southeast Australian Shelf
43 Southwest Australian Shelf
44 West-Central Australian Shelf
45 Northwest Australian Shelf
46 New Zealand Shelf
47 East China Sea
48 Yellow Sea
49 Kuro Shio Current
50 Sea of Japan
51 Oya Shio Current
52 Okhotsk Sea
53 West Bering Sea
54 Chukchi Sea
55 Beaufort Sea
56 East Siberian Sea
57 Laptev Sea
58 Kara Sea
59 Iceland Shelf
60 Faroe Plateau
61 Antarctic
62 Black Sea
63 Hudson Bay
64 Arctic Ocean

Estuaries

Located where freshwater from rivers flows into and mixes with saltwater from oceans, estuaries are vital ecosystems which provide tremendous natural, cultural and economic resources throughout the world. Estuaries include areas more commonly known as bays, sounds, inlets, and harbors. This transition zone between the land and ocean includes critical habitats such as salt marshes, shellfish beds, wetlands, mangroves, sea grass beds, and kelp and coral communities. These habitats provide important nursery areas for commercially valuable fish and shellfish, storm protection through buffering of wave and flooding events, and an important nutrient-filtering role that protects estuarine water quality.

Throughout the world estuaries are culturally important as, historically, civilizations established themselves along coastlines where food was abundant due to the highly productive nature of these habitats. This pattern continues with observed increased population densities in coastal areas, as estuaries not only provide resources for human consumption and employment (i.e. fisheries and ports) but recreational opportunities (i.e. tourism) and aesthetic value as well. Within the U.S., commercial and recreational fishing, boating, and tourism provide more than twenty-eight million jobs. Over 75 per cent of commercial fish rely on estuaries at some point in their life cycle and U.S. revenues from fishing activities alone are estimated to be greater than U.S.$111 billion annually. Worldwide, the economic benefits of productive estuarine systems to local communities are equally impressive.

Sustaining estuarine ecosystem functions and protecting human health under increasing coastal development pressures will continue to be a major challenge for decades to come. Advanced research, monitoring, and restoration tools, including remote sensing and in-situ rapid detection sensors, are used to support coastal management decisions with the goal of protecting and restoring estuaries to ensure that these unique ecosystems continue to provide their considerable benefits for future generations.

RIGHT:
In British Columbia, Canada, the fresh water of the Fraser River, clouded with sediment (right), meets the deep water of the Pacific Ocean (left).

FAR RIGHT:
Landsat satellite image of Chesapeake Bay on the northeastern coast of the U.S.A.. Water is blue, vegetation is green and urban areas are purple. Fields and other cultivated areas form a patchwork of pink and green (upper right). Brown coastal areas are marshes which support many thousands of water birds. Washington D.C. is the city at the top left of the image.

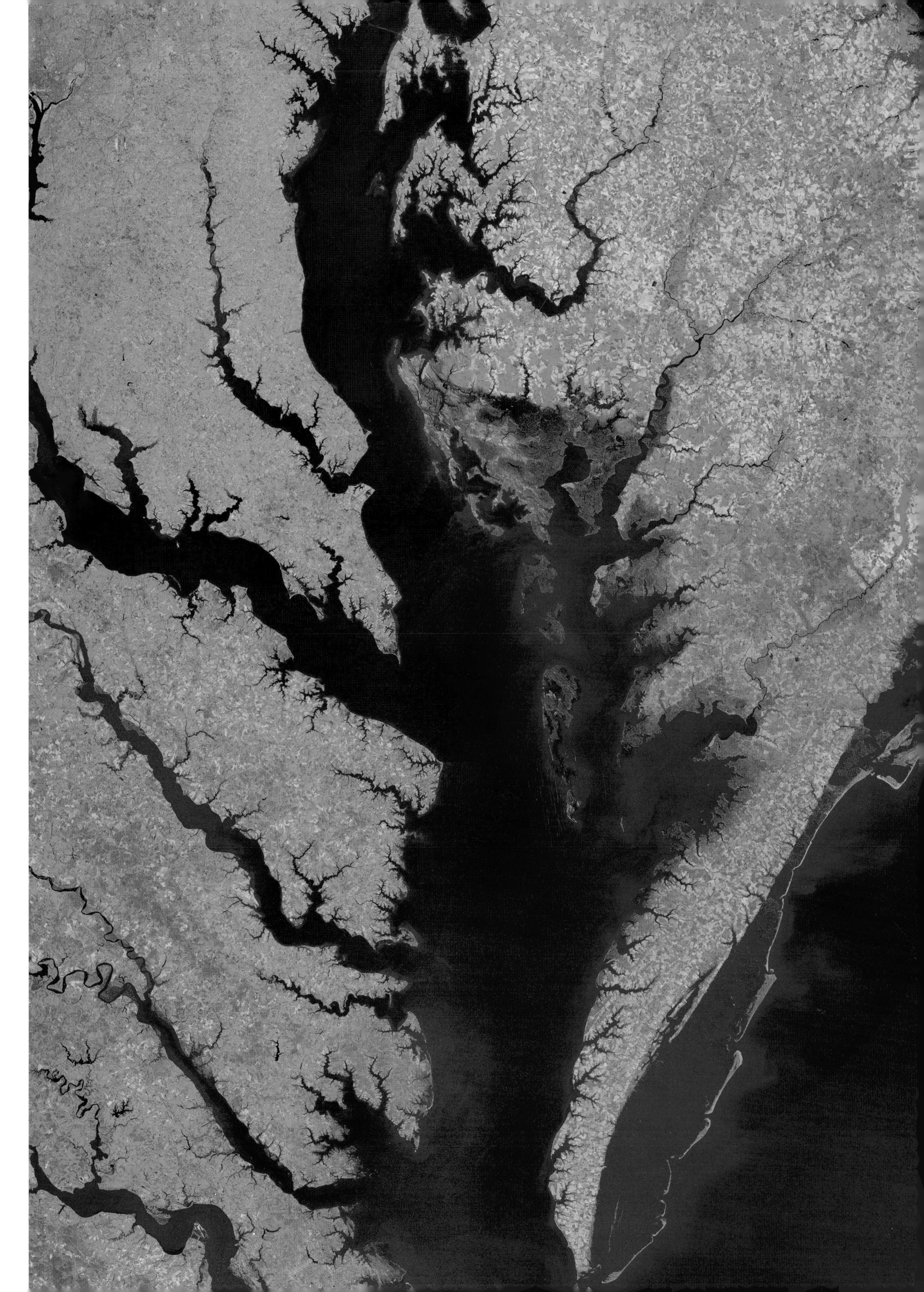

Wetlands

A wetland is an ecosystem which forms when soils are soaked by water long enough that the plants and animals must be adapted to flooded conditions and the micro-organisms in the soil can function with a lack of oxygen. The form wetlands take can vary greatly and these ecosystems can be found at almost all latitudes. Worldwide, there are roughly 2.2 million sq miles (5.6 million sq km) of wetlands, an area about four times the size of Alaska. For generations, people tended to view wetlands as wastelands which at best were annoyances and at worst places which bred disease. As we've come to understand these ecosystems better, the importance of these areas to both large-scale ecological function and human communities has become increasingly clear.

Wetlands are some of the earth's most productive ecosystems, providing food, nutrients, and oxygen which support surrounding ecosystems. They support fisheries, clean our waters, and lessen the impacts of storms and floods. Wetlands are also home to diverse plants and animals, including many threatened and endangered species. These habitats may also play a role in climate regulation through the extraction of carbon dioxide from the atmosphere.

Wetland loss worldwide has been extensive and the primary reasons for this loss are human related. Wetlands have been filled in to "reclaim" land for development or agriculture, water flow has been changed so that wetlands dry up or have a significant change in water quality, wetland products have been over-exploited, and new species have been introduced which fundamentally change the characteristics and function of wetland areas. In the 1600s, there were an estimated 220 million acres (89 million ha) of wetlands in the forty-eight contiguous States. By the mid 1980s, the wetlands in the same area had been reduced to 103.3 million acres (41.8 million ha). Worldwide, similar patterns exist although losses are higher in Europe, North America, and Asia than they are in South America and Africa, due probably to the differences in development and agriculture in those areas.

The news regarding wetlands isn't all bad, though. With the recognition of the value of wetlands, the loss of this habitat type is slowing in some regions. Efforts are going on worldwide to recreate and restore these critical areas and return both the functions and benefits these areas provide. In addition to restoration of degraded wetlands, existing ecosystems must continue to be conserved.

RIGHT:
Buttonbush is a wetland shrub that can grow up to 10 ft (3 m) tall and in water as deep as 4 ft (1.2 m). It provides habitat to invertebrates, who in turn are food for fish and wildlife.

ABOVE:
Battle Creek Swamp in Maryland, U.S.A.. Bald cypress once dominated extensive swamps, from Delaware southward and throughout much of the Gulf Coast. Many cypress swamps have been severely impacted by logging and development.

RIGHT:
Satellite photograph of the Florida Peninsula showing various coastal wetlands areas in dark green. The Everglades are on the southern tip of the state.

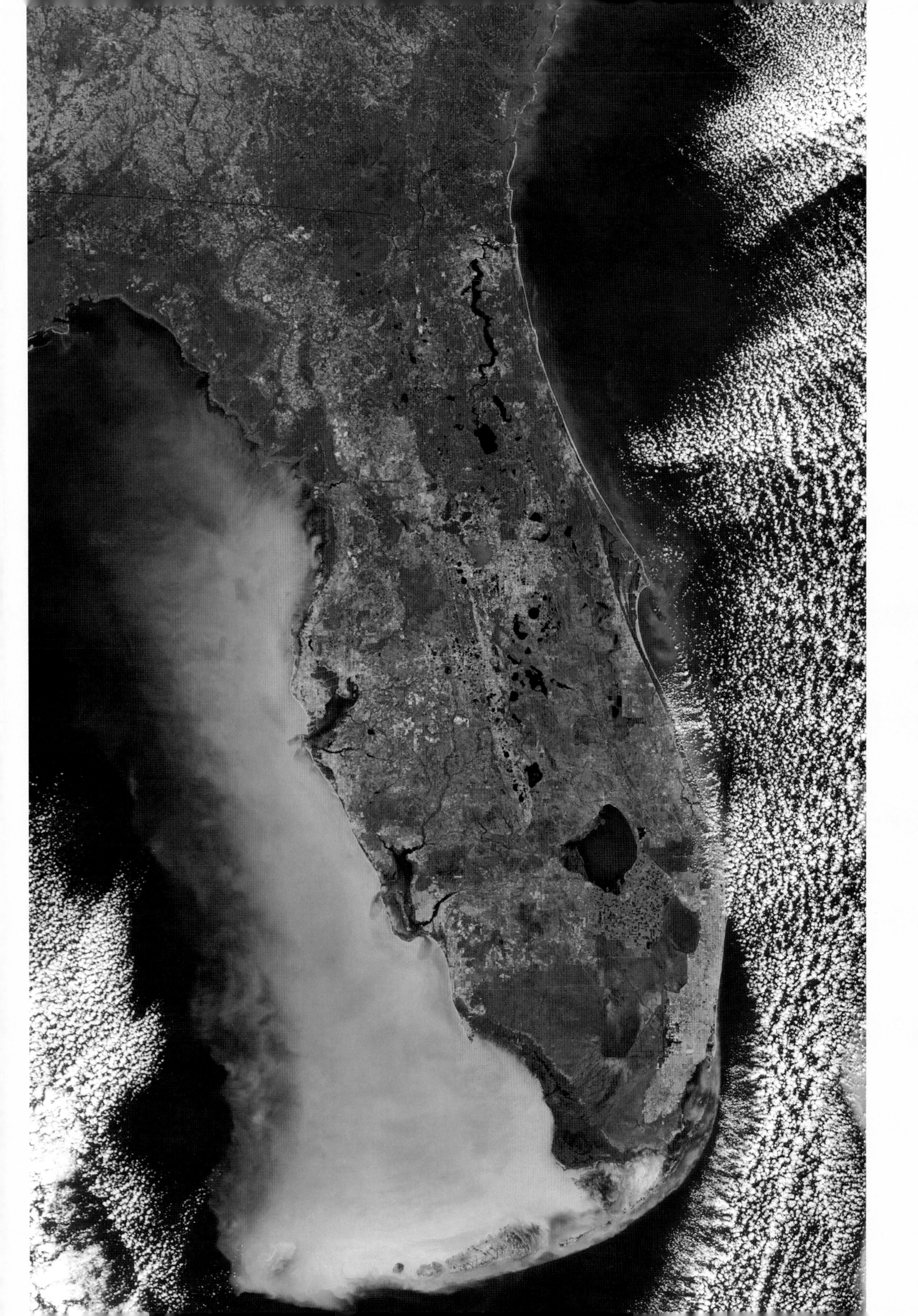

Salt marshes

Salt marshes are coastal wetlands which are flooded and drained by tides. Characterized by emergent soft-stemmed vegetation and saturated soils, they can be thought of as transition zones between land and water. Within these systems, gradients in salinity and elevation create a variety of microhabitats. Salinity can range from a high of 35 parts per thousand in saline marshes to 0.5 ppt in brackish marshes. Salt marshes occur worldwide, particularly in middle to high latitudes. Thriving along protected shorelines, they are a common habitat in estuaries. In the U.S., salt marshes can be found on every coast. Approximately half of the nation's salt marshes are located along the Gulf Coast.

Salt marshes are highly productive ecosystems supporting a diversity of crustaceans, mollusks, fish, birds, mammals, and reptiles. Common plants include grasses, sedges, and rushes. While some animals are salt marsh residents, others use the habitat periodically for feeding, spawning, or refuge. The plants and animals which inhabit salt marshes are adapted to the changing and sometimes harsh conditions.

Salt marshes are essential for healthy fisheries, coastlines, and communities. They are an integral part of our economy and our culture. These intertidal habitats provide essential food, refuge, or nursery habitat for more than 75 per cent of fisheries species, including shrimp, blue crab, and a variety of finfish. Salt marshes also protect shorelines from erosion by buffering wave action and trapping sediments. They reduce flooding by slowing and absorbing rainwater and protect water quality by filtering runoff, and by metabolizing excess nutrients.

Salt marshes support many endangered and threatened species, such as piping plovers and clapper rails and support such popular human activities as birding, hunting, fishing, and tourism.

ABOVE:
Salt marsh at mouth of Little River.

RIGHT:
The piping plover, which is found along the Atlantic coast of North America. Habitat loss and human activity near nesting sites have contributed to its decline in recent years. In 1986 it became a protected species under the Endangered Species Act.

RIGHT:
Marsh periwinkle, on marsh grass. Although it is not an endangered species it is an important part of the salt marsh community and can indicate the health of this habitat.

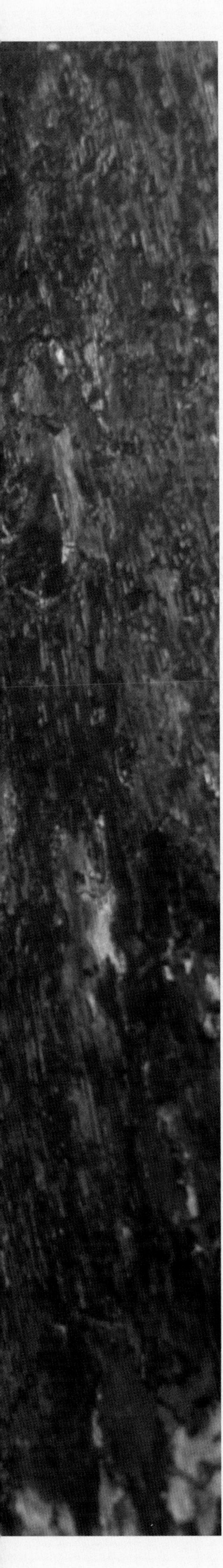

Mangroves

Mangroves are woody plants which provide an invaluable service to our coastal environment. They protect our coasts from storms and shoreline erosion, filter the water column, and support a diverse ecosystem which contributes to our economic and cultural environment. These plants are found primarily in brackish and saline areas within tropical and subtropical regions along the coast. The major mangrove species in the U.S., its territories, and the Caribbean are red mangroves, black mangroves, white mangroves, and buttonwood.

Mangroves roots and branches play a key role in preventing shoreline erosion by buffering and dispersing waves, filtering sediments and nutrients, and absorbing pollutants in the water column to maintain a healthy environment for plants, animals, and mangrove-associated habitats. Often, mammals, birds, fish, invertebrates, reptiles, and crustaceans use mangroves as breeding, feeding, or nursery grounds. For example, birds commonly nest and produce their young in high sections of mangrove trees and cover their eggs with the branches to protect juveniles from predators. Fish also feed within mangrove communities on benthic organisms, invertebrates, decayed leaves, fruits, and seeds floating in the water, and on bottom sediments. Typically, fish and crustaceans spawn amongst mangrove roots where nutrients are accessible. The roots and branches also protect juveniles from predation.

Mangroves are also important economically and culturally. For example, organisms such as finfish, bonefish, conch, lobster, and shrimp use mangroves as nursery grounds and eventually replenish fish stocks for household consumption as well as provide revenue for artisanal and commercial fishermen. Deterioration of mangrove health can cause reductions in fisheries biodiversity and productivity, lowering the country's seafood profits, and thus affecting the livelihoods of people who rely on mangrove ecosystems.

BELOW:
Mangrove roots

LEFT:
Underwater view of school of fish among mangrove trunks and branches at high tide, Cabo Delgado province, Mozambique.

Seagrass habitat

LEFT:
Neptune grass. This marine flowering plant is found only in the Mediterranean Sea. It is an important part of the ecosystem, as it provides habitat and food for many species. It is threatened by the accidental introduction of a tropical green alga in 1984.

Seagrass refers to forty-seven species of vascular plants which lead an almost totally rooted and submerged existence in shallow coastal marine systems. Worldwide, seagrass beds constitute one of the most common estuarine and coastal habitats. Few areas of the world's shallow coastal zone do not have seagrasses.

Research has demonstrated that seagrasses are important to coastal ecosystems. The plants and their epiphytes (plants which grow on another plant but are not parasitic) often contribute a large portion of the system's primary production. Marine turtles, wading birds, and fishes consume the leaves, many organisms graze on leaf epiphytes, and major food chains are seagrass detritus-based. Roots bind sediments, reducing erosion; leaves retard currents, increasing sedimentation. The habitat provides feeding grounds and refuge for fishes and other organisms (e.g. scallops, snails, crabs).

Man's dependence on estuarine and coastal environments for food production, energy development, transportation, waste disposal, and recreation is often incompatible with seagrasses. The ecological consequences of seagrass loss have been documented. For example, it has been estimated that the loss of 1,100 metric tonnes of seagrass in Florida resulted in the immediate loss of 1,800 metric tonnes of infauna, and 73 metric tonnes of fishery production.

The importance of these critical and sensitive habitats has led to efforts to protect them from destructive practices like dredging, trawling, and pollutant outfalls. Strategies to ameliorate seagrass losses have been examined. Studies which deal with restoring seagrasses and subsequent monitoring have been published widely. Restoration researchers have employed plants from healthy habitats, seeds, or natural recolonization but lack of regard to appropriate techniques and site selection resulted in many failures in early restoration efforts. Restoration is becoming more extensive and scientifically-based, and efforts take place on all U.S. coasts and in many countries including Australia, Spain, Portugal, and Japan, just to name a few.

BELOW:
Starfish at Olympic Coast National Marine Sanctuary. Different kinds of invertebrates, like this starfish, are commonly found in estuarine waters.

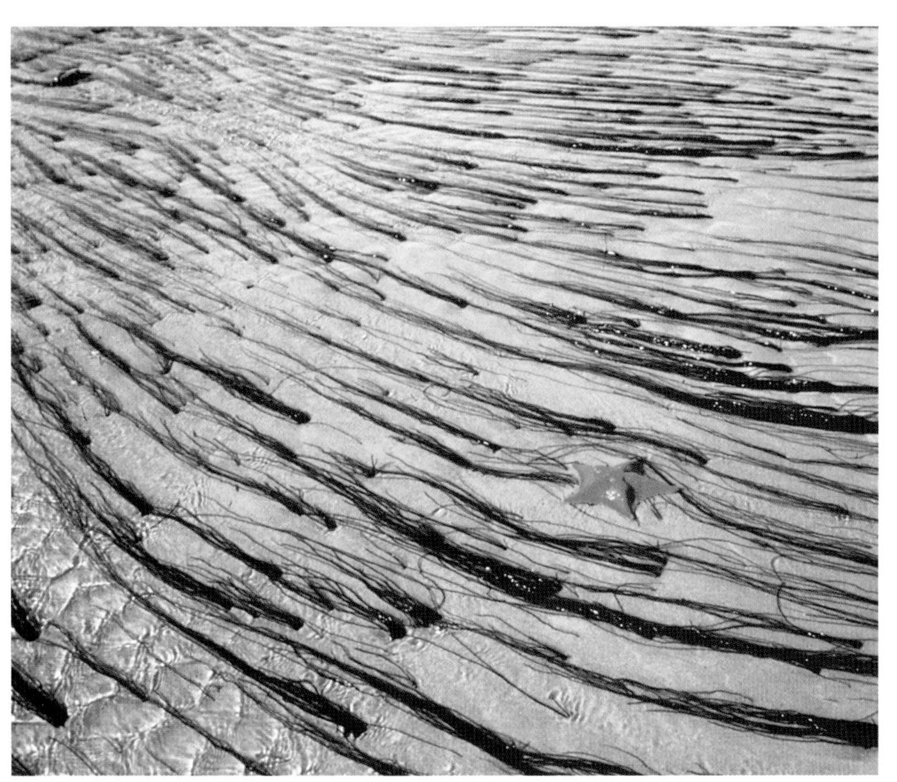

BELOW:
A NOAA researcher diving on a seagrass bed to assess the extent and condition of the bed.

Kelp forests

Kelp beds (or kelp forests) are nearshore shallow subtidal habitats dominated by large brown macroalgae of the order Laminariales. They occur in regions where hard substrate provides sufficient attachment places, where cool (<68° F /20°C) and nutrient-rich water and good light conditions prevail. Kelps attach to the rock surface with a holdfast, from which one or more stipes and one or many blades arise. Kelps have very fast growth rates and some species can reach 20-30 yards (18-27 m) in length within a few summer months. Some kelp species are only about a yard in length and are close to the bottom, creating the understory. Others extend all the way through the water column and reach the surface, forming a canopy (e.g. the giant kelp).

Kelps are so-called foundation organisms, which create complex, three-dimensional habitats which are used by a diverse array of other species. These organisms use kelps both for food and for shelter. Kelp-associated species live at various vertical strata in kelp forests: sea urchins, snails, seastars, limpets, chitons, abalone, worms, crustaceans, and small bottom fish are typical inhabitants close to the bottom. In mid water, several fish species use the structure afforded by kelps for protection or as foraging grounds. Several marine mammals (e.g. sea otters), birds, and some fish species have become specialized to live among the floating blades of canopy-forming kelp. In addition, the surface of kelp stipes and blades provides settlement areas for encrusting organisms such as bryozoans. Kelps are harvested and used by humans as food and food stabilizers. Kelp forests are under threat because of habitat destruction such as an increase of suspended inorganic sediments, which are especially detrimental to the microscopic overwintering stages of kelps. Climate change effects such as rising sea temperatures, increased storm activities, and reduced nutrient availability are also detrimental to kelp beds.

ABOVE:
Sea otter floating in a bed of kelp in southeastern Alaska, U.S.A.. Their streamlined bodies and webbed hind feet are ideally suited for hunting in water.

RIGHT:
Underwater forest of giant kelp, a type of brown algae, off the California coast, U.S.A.. A red coral is also seen at lower right.

Coral reef ecosystem

Corals are tiny animals, or polyps, which live in colonies of hundreds to thousands of individuals. There are two major types of corals: deep-water corals and shallow-water corals. Many deep-water corals lack the ability to secrete a stony skeleton and therefore do not form reef structures. Reef-building corals typically live in shallow tropical waters between 30° north and south of the equator, where sunlight can penetrate down to the reef.

A reef structure is created as coral polyps secrete a skeleton of Calcium carbonate ($CaCO_3$), or aragonite. Thousands of coral polyps make up one colony within a reef ecosystem, and these colonies together create larger reef structures from their aragonite skeletons. The shape and size of a reef structure depends on the species of corals in the colonies which make up the reef.

Reef-building corals have a symbiotic relationship with photosynthetic algae called zooxanthellae, which live within the tissues of the coral animal. Zooxanthellae provide the coral with oxygen and food through photosynthesis, and the coral provides the zooxanthellae with a protected environment. The foods zooxanthellae provide are essential for corals to build their skeletons.

Coral reefs and deep coral communities are some of the most diverse ecosystems on earth. Although they cover less than one per cent of the earth's surface, coral reefs provide habitat for hundreds of thousands of species of fish, invertebrates, and other marine organisms. Despite their ecological, economic, and cultural importance, coral reefs have significantly declined in recent decades due to a combination of human and natural stresses. Successful research, management, and conservation efforts will be critical to their future survival.

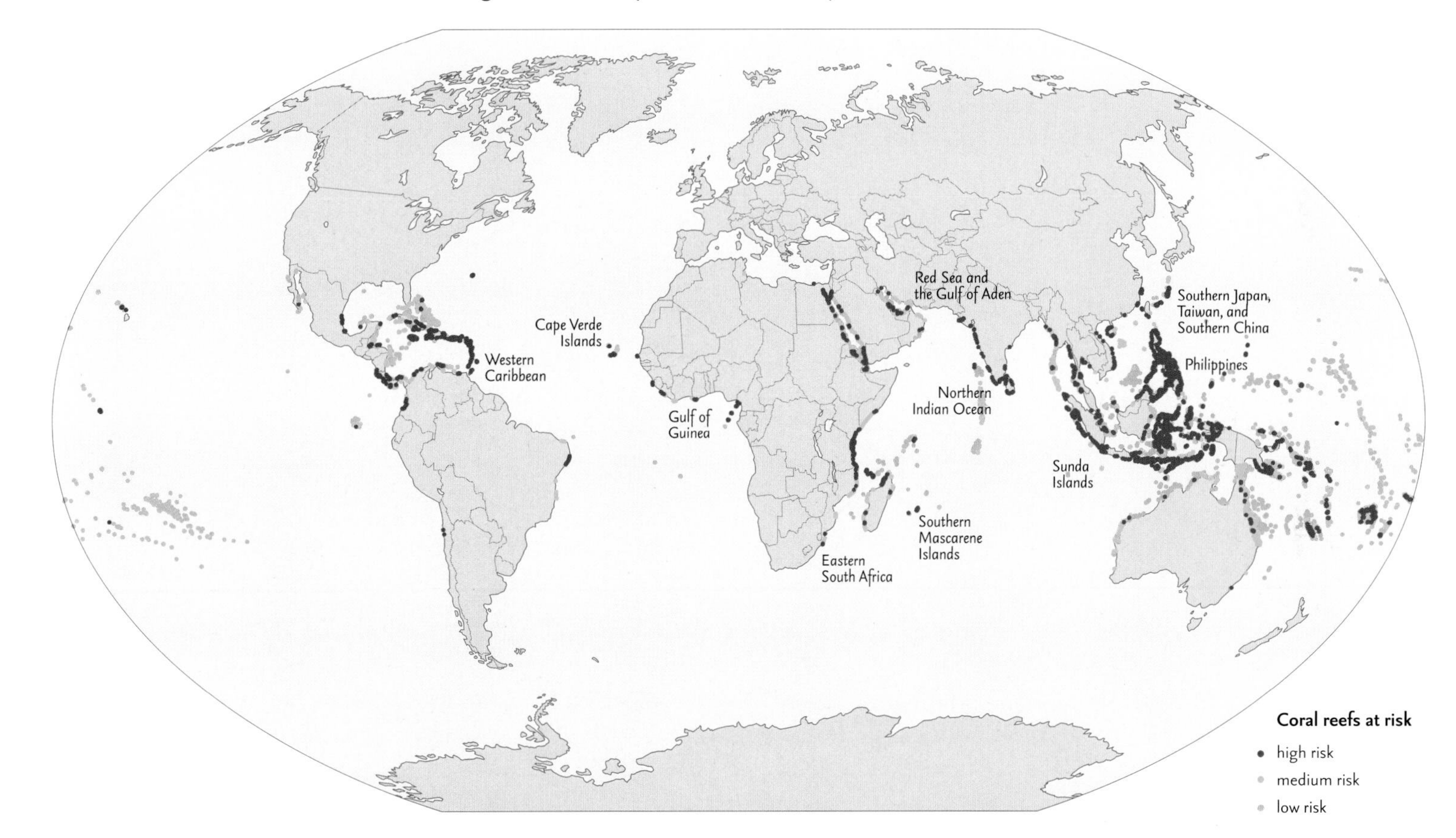

TOP LEFT:
Coral reefs provide habitat for a wide variety of fish and other organisms.

BOTTOM LEFT:
Lionfish on Rumitsuki Maru, Federated States of Micronesia.

RIGHT:
Lace coral on Hoyo Maru, Federated States of Micronesia.

LEFT:
Soft corals and tropical fish in the Mariana Arc region, western Pacific Ocean.

NEXT PAGE:
TOP ROW, LEFT TO RIGHT:
Green and purple anemone; Red jellyfish; Sea anemone with clown fish; Jellyfish and longfin bannerfish.

MIDDLE ROW, LEFT TO RIGHT:
Magnificent Star starfish; Sea urchin; Green sea turtle; Cabbage coral.

BOTTOM ROW, LEFT TO RIGHT:
Omilu or bluefin trevally; Featherduster worm; Colony of tunicates.

Exploring the oceans

Exploring the oceans
Introduction

Exploration is fundamental to the human spirit. Since the dawn of our species we have been explorers, motivated by everything from curiosity to the demands of survival and even spiritual inspiration. The ocean, our greatest resource yet to be explored, covers over 70 per cent of the earth's surface. And yet, only a fraction of its secrets are known.

Exploration of the ocean commenced with the first fishermen and navigators who used knowledge of the ocean for sustenance and trade. However, the scientific exploration of the ocean is a relatively recent phenomenon. Early sporadic attempts to plumb the depths and inventory the creatures of the sea were made, but the mid-nineteenth century saw the beginnings of the modern science of oceanography. Edward Forbes, a British naturalist, suggested that life was extinguished at 300 fathoms (550 m) water depth based on results from dredging cruises in the Aegean Sea. This theory, although wrong, was the first to inspire scientific exploration of the deep ocean. Exploring our continental shelves and slopes and the deep ocean for scientific purposes has gone on continuously since that time.

American and British scientists began scientific exploration of the sea after Forbes. The American Coast Survey began systematic Gulf Stream explorations in 1845; Matthew Fontaine Maury experimented with deep-ocean soundings beginning in 1849 which resulted in discovery of the Mid-Atlantic Ridge by 1853; and American and British surveyors ran the first complete line of soundings all the way across the Atlantic Ocean (from Ireland to Newfoundland) for a deep-sea telegraphic cable in 1856. This early work was followed by British, American, Swedish, and Norwegian investigations. Two milestones from this period are: the discovery of Josephine Bank by the Swedish corvette Josephine in 1869, which marked the first stand-alone seamount discovered as a result of intentional exploration; and the dredging up by HMS Porcupine of life from 2,435 fathoms (4,450 m) in 1869, exploding forever the concept of a lifeless ocean below 300 fathoms (549 m).

Many theories and discoveries have driven ocean exploration since that time. The search for the organism known as "Bathybius" and a great continuous "chalk" layer carpeting the seafloor were but two erroneous concepts which inspired the beginnings of the Challenger expedition. Bathybius proved to be an inanimate chemical precipitate; but the search for the chalk layer, an equally fallacious concept, led to discovering the nature and distribution of seafloor sediments. The Challenger expedition (1872–1876) subsequently proved to be the greatest of nineteenth-century exploratory voyages, circumnavigating the earth, and conducting a grand reconnaissance of the world ocean.

By the early twentieth century most of the large species residing in the ocean had been described, the basic configuration of the seafloor was outlined, and average conditions and movements of much of the water of the oceans had been determined. Yet, there were great discoveries and syntheses of knowledge still in store. Among the greatest was the development of the theory of plate tectonics which led to the shocking discovery of oceanic hydrothermal vent systems and chemosynthetic life.

Today's explorers are studying the interaction of ocean and atmosphere and its effect on climate, the inter-relationship between species and the physical conditions they require for life, the details of oceanic circulation, and, perhaps most important, the role humans have in altering the oceanic environment with the accompanying realization that we are stewards of that environment.

The ocean is vast and it is improbable that we will ever unlock all of its secrets. There will be explorers and scientists delving into its mysteries from the shore to the bottom of its deepest trenches for generations to come. They do it because they are curious, but also with a sense of urgency, as it is becoming increasingly apparent that the future of humanity and the future of the oceans are inextricably intertwined.

PREVIOUS PAGE:
Artist's conception of remotely operated vehicles (ROVs) exploring a black smoker vent field at Axial Volcano on the Juan de Fuca Ridge off the northwest coast of the United States.

RIGHT:
This map of the Atlantic Ocean was produced by Sir Wyville Thomson of the Challenger Expedition and was the first to show the continuity of the mid-Atlantic Ridge. The N-S line in the south Atlantic serendipitously followed the ridge crest and is among the most important ocean survey lines ever run as it led to the realization that the ridge was a continuous feature.

CONTOUR MAP OF THE ATLANTIC

From Soundings and Temperature Observations up to May, 1876.

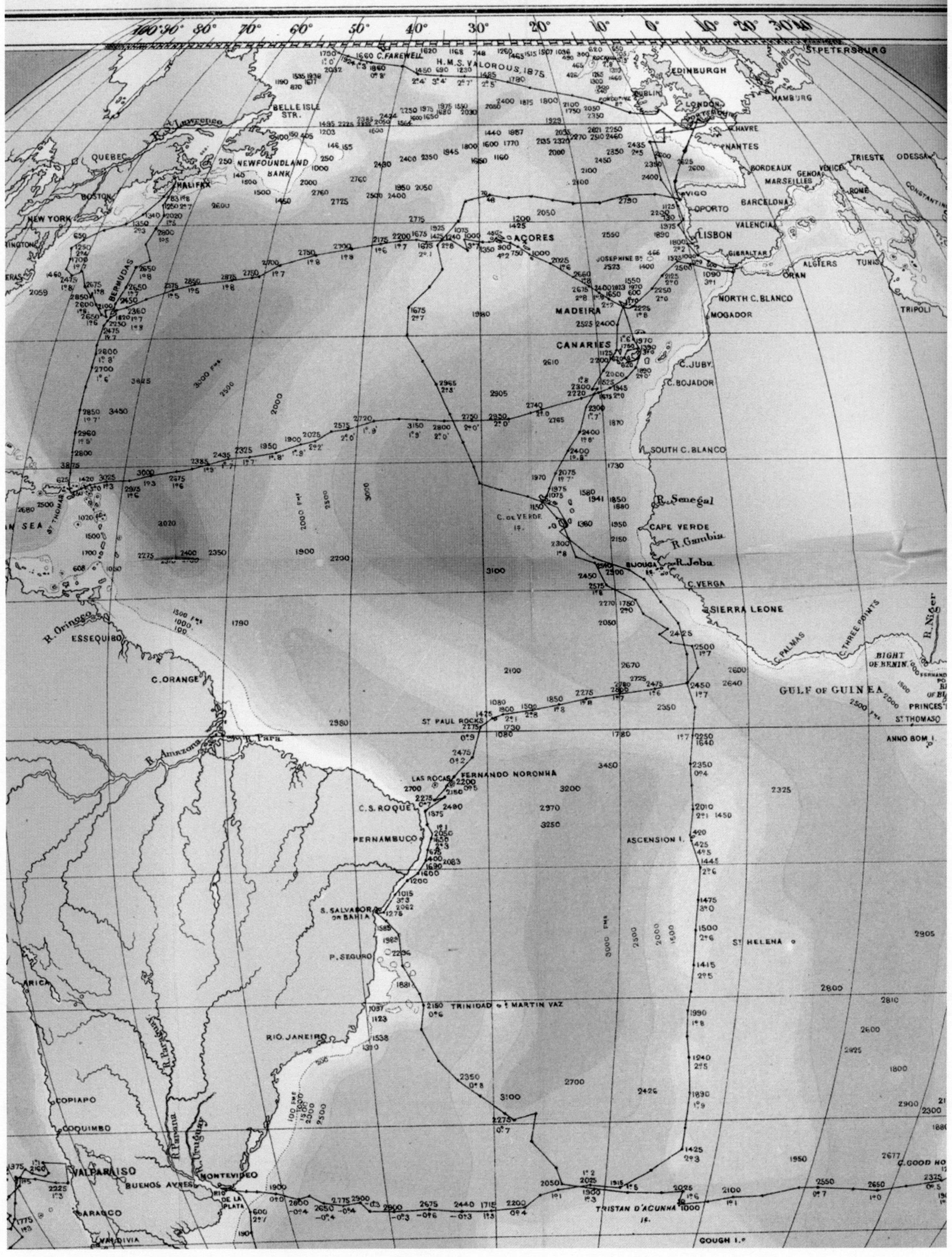

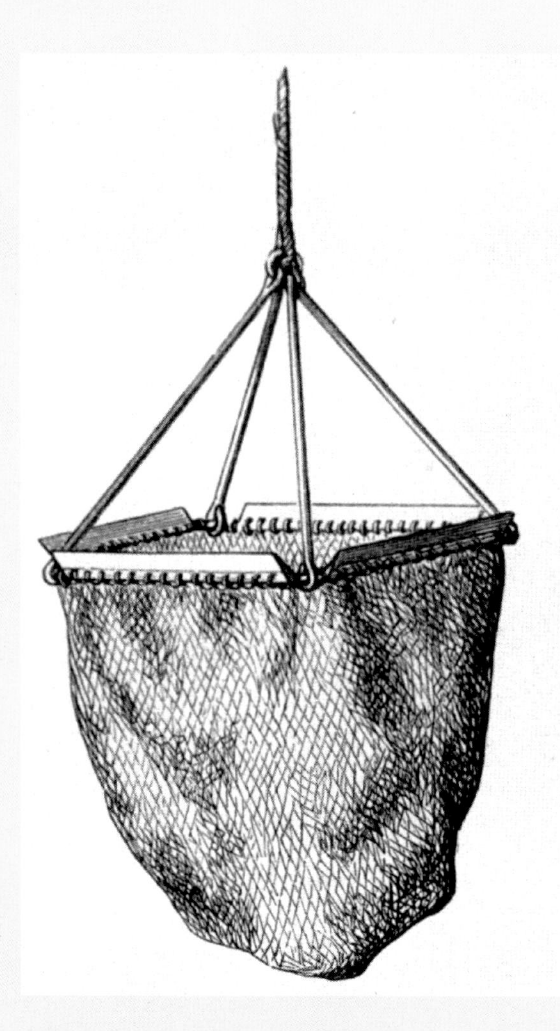

ABOVE LEFT:
Otho Frederick Müller's Dredge, 1750. Müller was the first person to dredge for purely scientific reasons.

ABOVE CENTER:
Drilled roller (round boulder) used as a sounding weight.

ABOVE RIGHT:
Ordinary sounding lead weight, dating from the Christian era.
The trapezoidal form, with a recess for tallow and acquiring bottom samples, dates from a few centuries back. In this form, millions of soundings were made.

LEFT:
Expedition of Christopher Columbus landing at Hispaniola, 1492.

RIGHT:
Chart of the Gulf Stream by Benjamin Franklin, dating from around 1782.

Early exploration and discovery

Up until the beginnings of European exploration of our world, there was very little knowledge of the science which today we call oceanography. What knowledge existed was acquired as the result of the practical needs of navigators and fishermen. A notable exception was Aristotle's work in marine biology in the late fourth century BC. In the first century BC a 1,000-fathom (1,829 m) sounding into the Tyrrhenian Sea was reported on by the geographer Strabo, but this remained humanity's sole foray into the deep sea for nearly 2,000 years.

Few advances were made until the European Age of Exploration. Using magnetic compass, astrolabe, and other crude navigational instruments, Prince Henry the Navigator of Portugal sent ships down the coast of Africa in the first large-scale project of systematic geographic exploration. Climatic conditions and previously unknown oceanic currents were among the discoveries. Columbus, sailing for Spain, made a number of scientific discoveries including noting the westward flowing limb of the north Atlantic gyre, the change to northeasterly magnetic declination when crossing the Atlantic, and the Sargasso Sea. He even attempted to sound in the north Atlantic with 200 fathoms (365 m) of line for a first modern attempt at deep-ocean sounding.

In the late seventeenth century, English scientists studied the temperature and "saltness" of oceanic waters. In 1725 the Italian nobleman Luigi Marsigli produced the first treatise on oceanography, Histoire physique de la mer. Although the pace accelerated in the eighteenth century with the development of means to determine longitude at sea, the publication of a first map of the Gulf Stream by Benjamin Franklin, scientific dredging work by Otho Frederick Müller, and the voyages of Captain James Cook, very little effort was devoted to oceanography. It remained until the mid-nineteenth century for oceanography to blossom as a recognized scientific discipline.

Portolans, nautical charts, and mapping the ocean

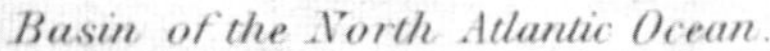

In the late thirteenth century Europe was emerging from the Dark Ages. During this period there suddenly appeared from the hand of some unknown genius, a first rational look at the world around us. This look was in the form of the portolan chart, a chart which showed the world the way it is, as opposed to a geography of the imagination based on superstition. The earliest surviving portolan chart is dated 1290. Although the portolan style charts displayed considerable geographic knowledge of shoreline and ports, not one depth was plotted on any of them.

In 1543, a single depth written in script was recorded on a map of the Gulf of California. Then in 1584, Lucas Janszoon Waghenaer invented the modern nautical chart showing soundings written as Arabic numerals, various hazards to navigation including shoals and rocks, a scale, and compass rose. This was but a small step forward in understanding the oceans; charting efforts were confined to coastal regions as the deep ocean was considered "unfathomable."

It was another 200 years before sporadic efforts were made to sound the deep sea, beginning with the British ship HMS Racehorse which sounded in 683 fathoms (1,250 m) in the Norwegian Sea in 1773. Sixty-seven years later Sir James Clark Ross sounded in the south Atlantic and found 2,425 fathoms (4,435 m), considered to be the first modern deep-sea sounding. By 1853 sufficient data had been acquired by British and American vessels for Matthew Fontaine Maury to construct a bathymetric map of the north Atlantic. This first map had many inaccuracies, but by the late nineteenth century exploratory voyages from many nations had sounded the ocean sufficiently to delineate most of its major features. Today, seafloor mappers are able to obtain unprecedented views of even very small features by use of data from acoustic systems which are processed by modern computers.

BELOW:
First map of an oceanic basin. Matthew Fontaine Maury produced this map of the north Atlantic Ocean in 1853 and published it in the Wind and Current Charts for that year.

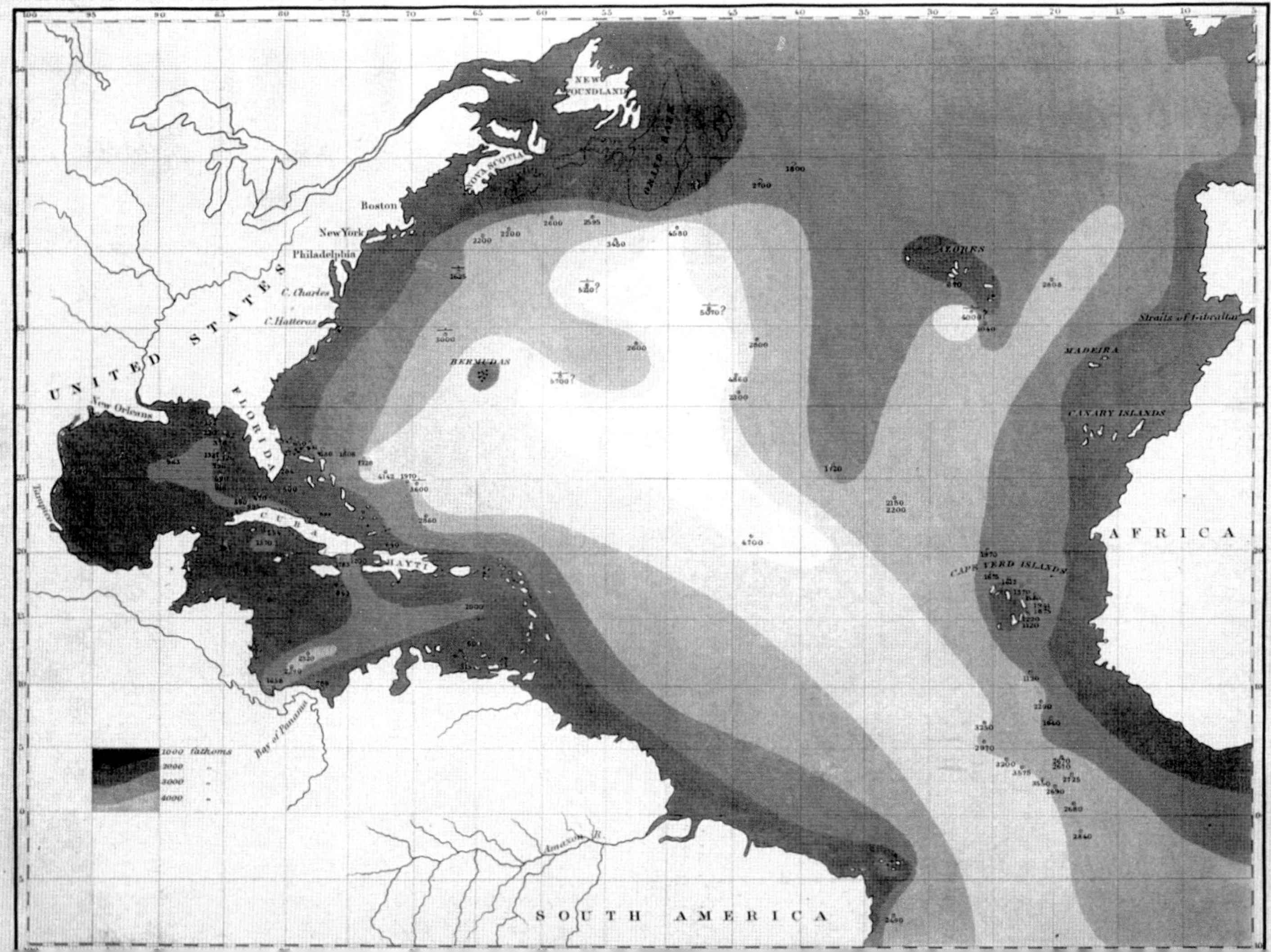

LEFT:
Captain Constantine Phipps' ship, the HMS Racehorse, in pack ice on September 4, 1773. It sounded in 683 fathoms (1,250 m) in the Norwegian Sea, the first modern successful sounding on a continental slope area.

BELOW:
Sounding technique used by Sir James Clark Ross in obtaining the first modern deep sea sounding on January 3, 1840 at latitude 27°26' S, longitude 17°29' W. The observed depth was 2,425 fathoms (4,435 m). On January 22, 1968, the ESSA ship Discoverer sounded at the same location with a modern echo-sounder and measured 2,312 fathoms (4,229 m), less than a 5 per cent error for Ross's observation.

The legacy of the Challenger expedition

No voyage in the history of ocean exploration is more famous than that of HMS Challenger. The Challenger sailed over 68,000 miles (109,430 km) and circumnavigated the earth between December 1872 and May 1876. Although there had been many previous scientific expeditions, this marked the first circumnavigation for the sole purpose of studying the ocean. Oceanography was developing as a science and many nations were involved, but the Challenger expedition stands out as a major landmark. It was the first grand reconnaissance of all the major ocean basins. During the cruise fundamental discoveries were made which reverberate to this day, including the recognition of the continuity of the Mid-Atlantic Ridge, the composition and distribution of many types of seafloor sediments, the similarity of many life-forms throughout the oceanic basins, and early indications of distinct water masses driving the deep circulation of the ocean waters.

In a larger sense, the major contribution of the Challenger expedition was the grand synthesis of oceanic knowledge brought together by Wyville Thomson and John Murray following the completion of the voyage. Thomson began the work of producing the fifty volumes of the Challenger Reports and after his death in 1882, this task fell to John Murray. The Reports used the work and collections of the expedition as the foundation, but drew on the talents and knowledge of scientists from many nations to analyze the Challenger collections while also incorporating the results of other expeditions into the final volumes. It is to the credit of Thomson and Murray that they comprehended that the ocean belonged to all and that understanding it required the work of the best scientists regardless of nationality. This tradition of international collaboration for studying the oceans continues to this day.

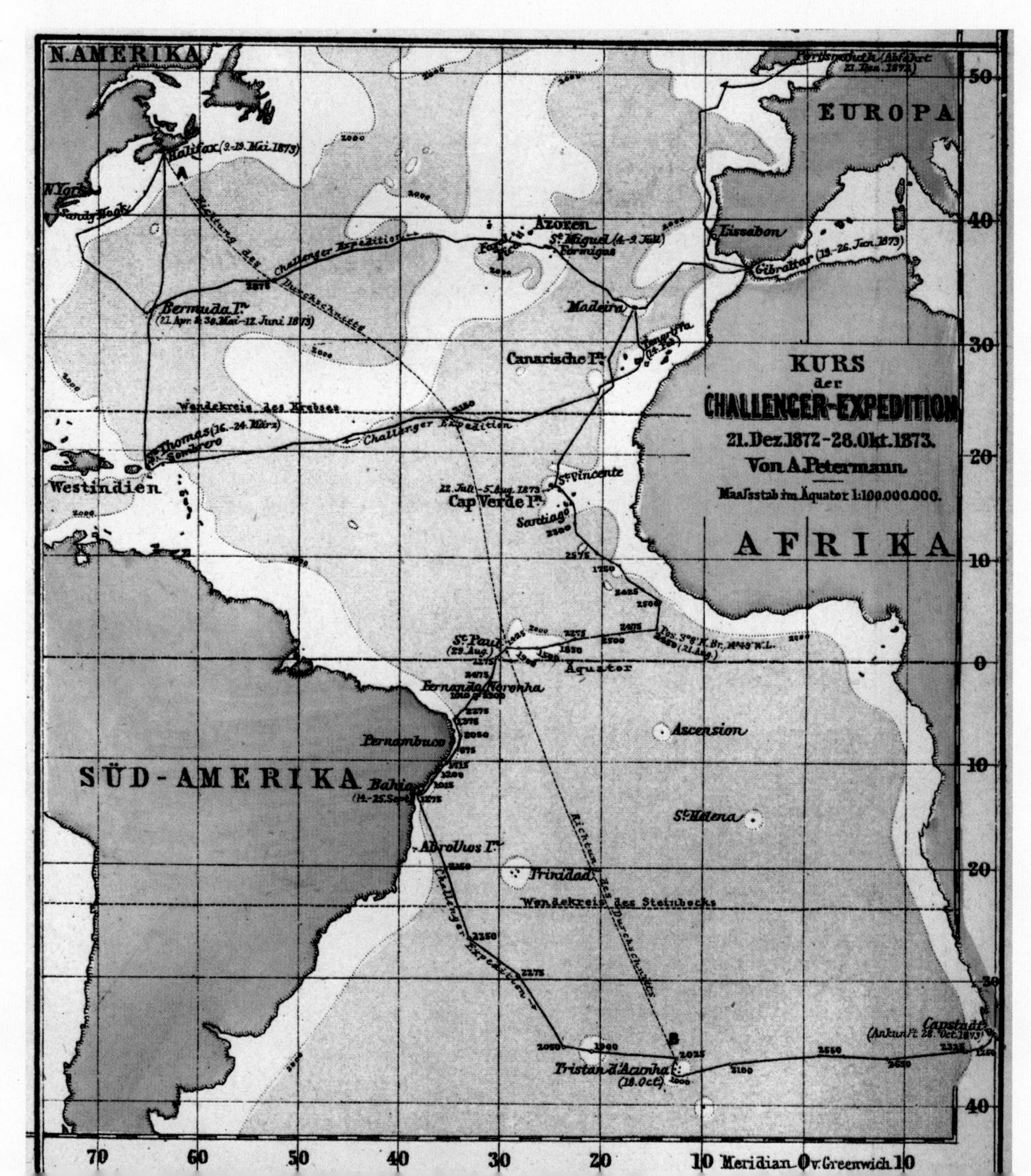

RIGHT:
Map of the Atlantic Ocean published after the outward bound segment of the Challenger expedition. This map shows little improvement over earlier versions. The Mid-Atlantic Ridge is shown bending to the west, south of the Azores and there is no expression of the ridge in the south Atlantic Ocean.

OPPOSITE PAGE TOP:
HMS Challenger, from "The Voyage of the Challenger – The Atlantic" Vol I, by Sir Wyville Thomson, 1878.

OPPOSITE PAGE BOTTOM LEFT:
Trawl, dredge, tow-net, water-bottle, sounding machine, and sieves used by HMS Challenger.

OPPOSITE PAGE BOTTOM RIGHT:
Sir John Murray of HMS Challenger. The photo was taken c. 1900.

Ocean exploration timeline

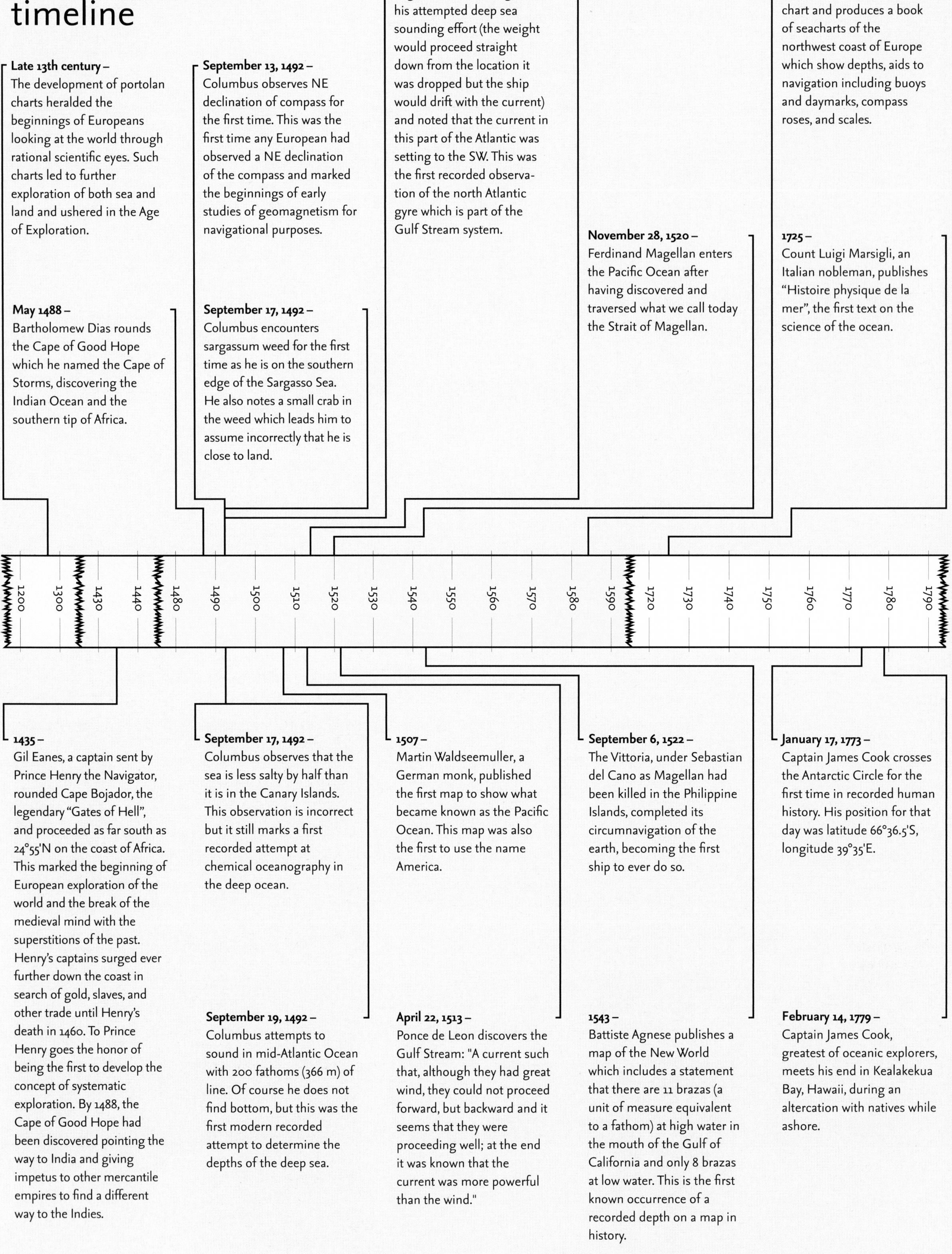

Late 13th century –
The development of portolan charts heralded the beginnings of Europeans looking at the world through rational scientific eyes. Such charts led to further exploration of both sea and land and ushered in the Age of Exploration.

September 13, 1492 –
Columbus observes NE declination of compass for the first time. This was the first time any European had observed a NE declination of the compass and marked the beginnings of early studies of geomagnetism for navigational purposes.

September 19, 1492 –
Columbus observes the angle of the sounding line in his attempted deep sea sounding effort (the weight would proceed straight down from the location it was dropped but the ship would drift with the current) and noted that the current in this part of the Atlantic was setting to the SW. This was the first recorded observation of the north Atlantic gyre which is part of the Gulf Stream system.

September 26, 1513 –
Balboa discovers the Pacific Ocean.

1584 –
Lucas Janszoon Waghenaer invents the modern nautical chart and produces a book of seacharts of the northwest coast of Europe which show depths, aids to navigation including buoys and daymarks, compass roses, and scales.

May 1488 –
Bartholomew Dias rounds the Cape of Good Hope which he named the Cape of Storms, discovering the Indian Ocean and the southern tip of Africa.

September 17, 1492 –
Columbus encounters sargassum weed for the first time as he is on the southern edge of the Sargasso Sea. He also notes a small crab in the weed which leads him to assume incorrectly that he is close to land.

November 28, 1520 –
Ferdinand Magellan enters the Pacific Ocean after having discovered and traversed what we call today the Strait of Magellan.

1725 –
Count Luigi Marsigli, an Italian nobleman, publishes “Histoire physique de la mer”, the first text on the science of the ocean.

1435 –
Gil Eanes, a captain sent by Prince Henry the Navigator, rounded Cape Bojador, the legendary “Gates of Hell”, and proceeded as far south as 24°55'N on the coast of Africa. This marked the beginning of European exploration of the world and the break of the medieval mind with the superstitions of the past. Henry’s captains surged ever further down the coast in search of gold, slaves, and other trade until Henry’s death in 1460. To Prince Henry goes the honor of being the first to develop the concept of systematic exploration. By 1488, the Cape of Good Hope had been discovered pointing the way to India and giving impetus to other mercantile empires to find a different way to the Indies.

September 17, 1492 –
Columbus observes that the sea is less salty by half than it is in the Canary Islands. This observation is incorrect but it still marks a first recorded attempt at chemical oceanography in the deep ocean.

1507 –
Martin Waldseemuller, a German monk, published the first map to show what became known as the Pacific Ocean. This map was also the first to use the name America.

September 6, 1522 –
The Vittoria, under Sebastian del Cano as Magellan had been killed in the Philippine Islands, completed its circumnavigation of the earth, becoming the first ship to ever do so.

January 17, 1773 –
Captain James Cook crosses the Antarctic Circle for the first time in recorded human history. His position for that day was latitude 66°36.5'S, longitude 39°35'E.

September 19, 1492 –
Columbus attempts to sound in mid-Atlantic Ocean with 200 fathoms (366 m) of line. Of course he does not find bottom, but this was the first modern recorded attempt to determine the depths of the deep sea.

April 22, 1513 –
Ponce de Leon discovers the Gulf Stream: "A current such that, although they had great wind, they could not proceed forward, but backward and it seems that they were proceeding well; at the end it was known that the current was more powerful than the wind."

1543 –
Battiste Agnese publishes a map of the New World which includes a statement that there are 11 brazas (a unit of measure equivalent to a fathom) at high water in the mouth of the Gulf of California and only 8 brazas at low water. This is the first known occurrence of a recorded depth on a map in history.

February 14, 1779 –
Captain James Cook, greatest of oceanic explorers, meets his end in Kealakekua Bay, Hawaii, during an altercation with natives while ashore.

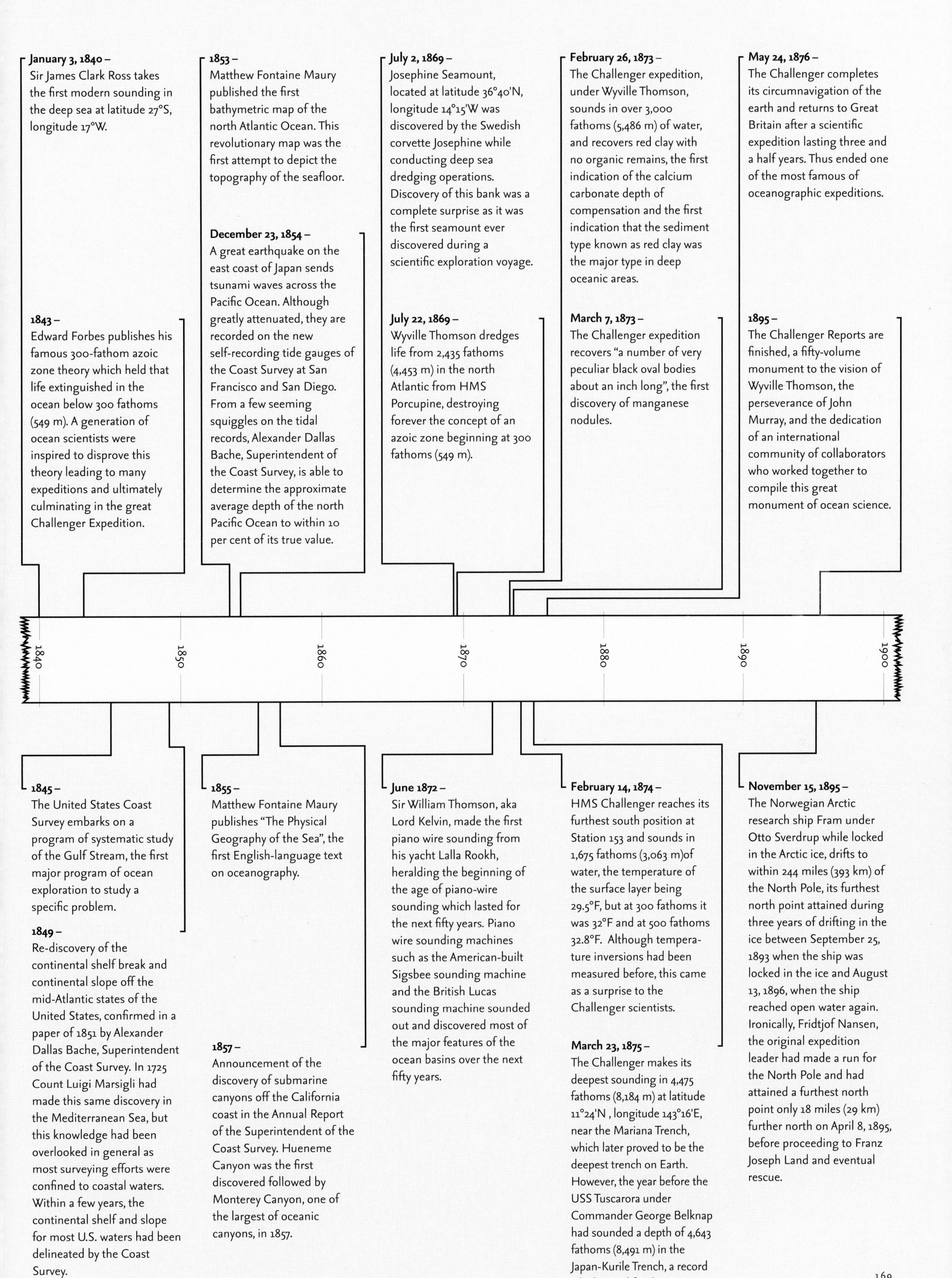

January 3, 1840 –
Sir James Clark Ross takes the first modern sounding in the deep sea at latitude 27°S, longitude 17°W.

1843 –
Edward Forbes publishes his famous 300-fathom azoic zone theory which held that life extinguished in the ocean below 300 fathoms (549 m). A generation of ocean scientists were inspired to disprove this theory leading to many expeditions and ultimately culminating in the great Challenger Expedition.

1853 –
Matthew Fontaine Maury published the first bathymetric map of the north Atlantic Ocean. This revolutionary map was the first attempt to depict the topography of the seafloor.

December 23, 1854 –
A great earthquake on the east coast of Japan sends tsunami waves across the Pacific Ocean. Although greatly attenuated, they are recorded on the new self-recording tide gauges of the Coast Survey at San Francisco and San Diego. From a few seeming squiggles on the tidal records, Alexander Dallas Bache, Superintendent of the Coast Survey, is able to determine the approximate average depth of the north Pacific Ocean to within 10 per cent of its true value.

July 2, 1869 –
Josephine Seamount, located at latitude 36°40'N, longitude 14°15'W was discovered by the Swedish corvette Josephine while conducting deep sea dredging operations. Discovery of this bank was a complete surprise as it was the first seamount ever discovered during a scientific exploration voyage.

July 22, 1869 –
Wyville Thomson dredges life from 2,435 fathoms (4,453 m) in the north Atlantic from HMS Porcupine, destroying forever the concept of an azoic zone beginning at 300 fathoms (549 m).

February 26, 1873 –
The Challenger expedition, under Wyville Thomson, sounds in over 3,000 fathoms (5,486 m) of water, and recovers red clay with no organic remains, the first indication of the calcium carbonate depth of compensation and the first indication that the sediment type known as red clay was the major type in deep oceanic areas.

March 7, 1873 –
The Challenger expedition recovers "a number of very peculiar black oval bodies about an inch long", the first discovery of manganese nodules.

May 24, 1876 –
The Challenger completes its circumnavigation of the earth and returns to Great Britain after a scientific expedition lasting three and a half years. Thus ended one of the most famous of oceanographic expeditions.

1895 –
The Challenger Reports are finished, a fifty-volume monument to the vision of Wyville Thomson, the perseverance of John Murray, and the dedication of an international community of collaborators who worked together to compile this great monument of ocean science.

1840
1850
1860
1870
1880
1890
1900

1845 –
The United States Coast Survey embarks on a program of systematic study of the Gulf Stream, the first major program of ocean exploration to study a specific problem.

1849 –
Re-discovery of the continental shelf break and continental slope off the mid-Atlantic states of the United States, confirmed in a paper of 1851 by Alexander Dallas Bache, Superintendent of the Coast Survey. In 1725 Count Luigi Marsigli had made this same discovery in the Mediterranean Sea, but this knowledge had been overlooked in general as most surveying efforts were confined to coastal waters. Within a few years, the continental shelf and slope for most U.S. waters had been delineated by the Coast Survey.

1855 –
Matthew Fontaine Maury publishes "The Physical Geography of the Sea", the first English-language text on oceanography.

1857 –
Announcement of the discovery of submarine canyons off the California coast in the Annual Report of the Superintendent of the Coast Survey. Hueneme Canyon was the first discovered followed by Monterey Canyon, one of the largest of oceanic canyons, in 1857.

June 1872 –
Sir William Thomson, aka Lord Kelvin, made the first piano wire sounding from his yacht Lalla Rookh, heralding the beginning of the age of piano-wire sounding which lasted for the next fifty years. Piano wire sounding machines such as the American-built Sigsbee sounding machine and the British Lucas sounding machine sounded out and discovered most of the major features of the ocean basins over the next fifty years.

February 14, 1874 –
HMS Challenger reaches its furthest south position at Station 153 and sounds in 1,675 fathoms (3,063 m)of water, the temperature of the surface layer being 29.5°F, but at 300 fathoms it was 32°F and at 500 fathoms 32.8°F. Although temperature inversions had been measured before, this came as a surprise to the Challenger scientists.

March 23, 1875 –
The Challenger makes its deepest sounding in 4,475 fathoms (8,184 m) at latitude 11°24'N , longitude 143°16'E, near the Mariana Trench, which later proved to be the deepest trench on Earth. However, the year before the USS Tuscarora under Commander George Belknap had sounded a depth of 4,643 fathoms (8,491 m) in the Japan-Kurile Trench, a record which stood for the next twenty years.

November 15, 1895 –
The Norwegian Arctic research ship Fram under Otto Sverdrup while locked in the Arctic ice, drifts to within 244 miles (393 km) of the North Pole, its furthest north point attained during three years of drifting in the ice between September 25, 1893 when the ship was locked in the ice and August 13, 1896, when the ship reached open water again. Ironically, Fridtjof Nansen, the original expedition leader had made a run for the North Pole and had attained a furthest north point only 18 miles (29 km) further north on April 8, 1895, before proceeding to Franz Joseph Land and eventual rescue.

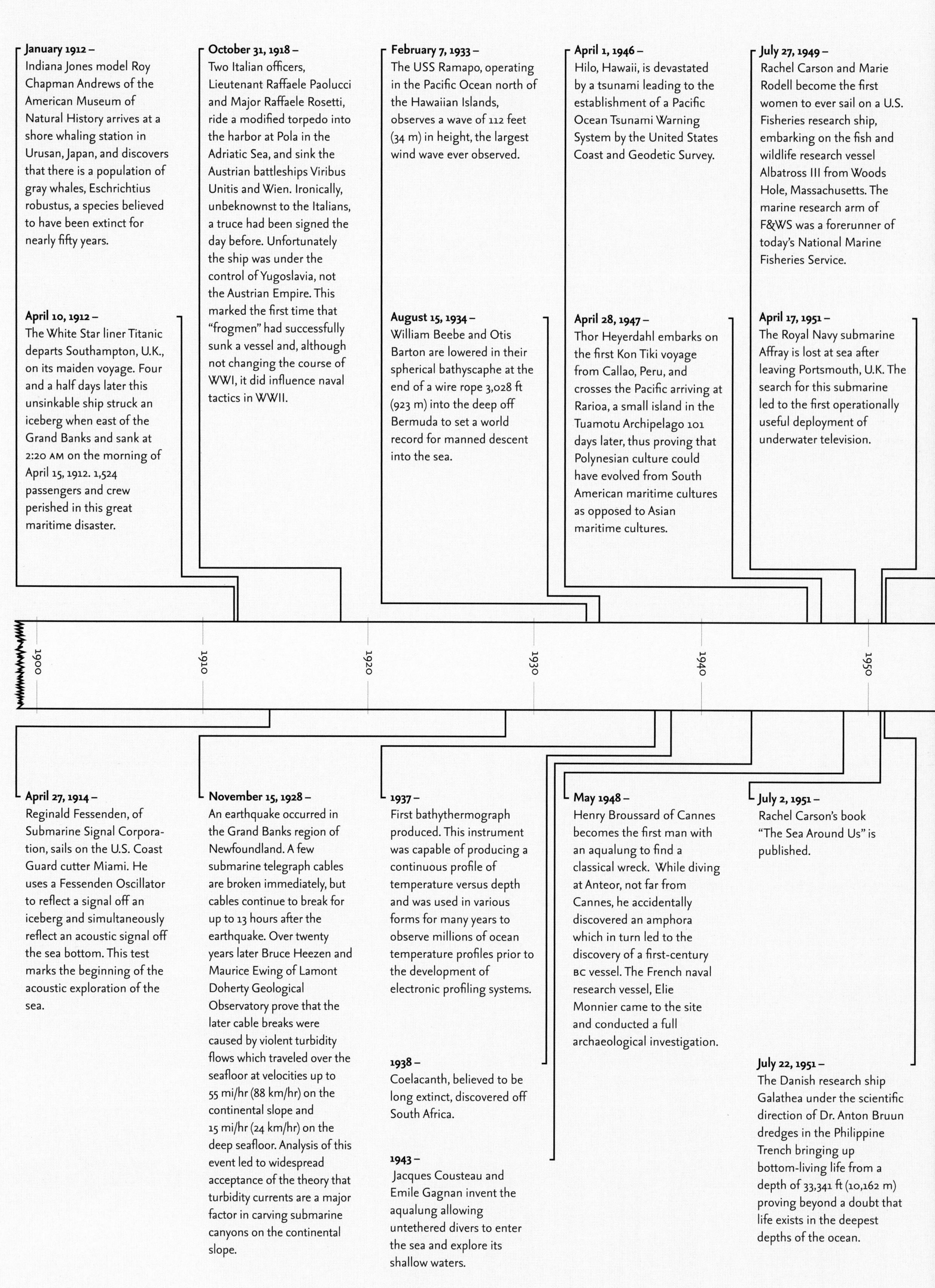

January 1912 –
Indiana Jones model Roy Chapman Andrews of the American Museum of Natural History arrives at a shore whaling station in Urusan, Japan, and discovers that there is a population of gray whales, Eschrichtius robustus, a species believed to have been extinct for nearly fifty years.

April 10, 1912 –
The White Star liner Titanic departs Southampton, U.K., on its maiden voyage. Four and a half days later this unsinkable ship struck an iceberg when east of the Grand Banks and sank at 2:20 AM on the morning of April 15, 1912. 1,524 passengers and crew perished in this great maritime disaster.

April 27, 1914 –
Reginald Fessenden, of Submarine Signal Corporation, sails on the U.S. Coast Guard cutter Miami. He uses a Fessenden Oscillator to reflect a signal off an iceberg and simultaneously reflect an acoustic signal off the sea bottom. This test marks the beginning of the acoustic exploration of the sea.

October 31, 1918 –
Two Italian officers, Lieutenant Raffaele Paolucci and Major Raffaele Rosetti, ride a modified torpedo into the harbor at Pola in the Adriatic Sea, and sink the Austrian battleships Viribus Unitis and Wien. Ironically, unbeknownst to the Italians, a truce had been signed the day before. Unfortunately the ship was under the control of Yugoslavia, not the Austrian Empire. This marked the first time that "frogmen" had successfully sunk a vessel and, although not changing the course of WWI, it did influence naval tactics in WWII.

November 15, 1928 –
An earthquake occurred in the Grand Banks region of Newfoundland. A few submarine telegraph cables are broken immediately, but cables continue to break for up to 13 hours after the earthquake. Over twenty years later Bruce Heezen and Maurice Ewing of Lamont Doherty Geological Observatory prove that the later cable breaks were caused by violent turbidity flows which traveled over the seafloor at velocities up to 55 mi/hr (88 km/hr) on the continental slope and 15 mi/hr (24 km/hr) on the deep seafloor. Analysis of this event led to widespread acceptance of the theory that turbidity currents are a major factor in carving submarine canyons on the continental slope.

February 7, 1933 –
The USS Ramapo, operating in the Pacific Ocean north of the Hawaiian Islands, observes a wave of 112 feet (34 m) in height, the largest wind wave ever observed.

August 15, 1934 –
William Beebe and Otis Barton are lowered in their spherical bathyscaphe at the end of a wire rope 3,028 ft (923 m) into the deep off Bermuda to set a world record for manned descent into the sea.

1937 –
First bathythermograph produced. This instrument was capable of producing a continuous profile of temperature versus depth and was used in various forms for many years to observe millions of ocean temperature profiles prior to the development of electronic profiling systems.

1938 –
Coelacanth, believed to be long extinct, discovered off South Africa.

1943 –
Jacques Cousteau and Emile Gagnan invent the aqualung allowing untethered divers to enter the sea and explore its shallow waters.

April 1, 1946 –
Hilo, Hawaii, is devastated by a tsunami leading to the establishment of a Pacific Ocean Tsunami Warning System by the United States Coast and Geodetic Survey.

April 28, 1947 –
Thor Heyerdahl embarks on the first Kon Tiki voyage from Callao, Peru, and crosses the Pacific arriving at Rarioa, a small island in the Tuamotu Archipelago 101 days later, thus proving that Polynesian culture could have evolved from South American maritime cultures as opposed to Asian maritime cultures.

May 1948 –
Henry Broussard of Cannes becomes the first man with an aqualung to find a classical wreck. While diving at Anteor, not far from Cannes, he accidentally discovered an amphora which in turn led to the discovery of a first-century BC vessel. The French naval research vessel, Elie Monnier came to the site and conducted a full archaeological investigation.

July 27, 1949 –
Rachel Carson and Marie Rodell become the first women to ever sail on a U.S. Fisheries research ship, embarking on the fish and wildlife research vessel Albatross III from Woods Hole, Massachusetts. The marine research arm of F&WS was a forerunner of today's National Marine Fisheries Service.

April 17, 1951 –
The Royal Navy submarine Affray is lost at sea after leaving Portsmouth, U.K. The search for this submarine led to the first operationally useful deployment of underwater television.

July 2, 1951 –
Rachel Carson's book "The Sea Around Us" is published.

July 22, 1951 –
The Danish research ship Galathea under the scientific direction of Dr. Anton Bruun dredges in the Philippine Trench bringing up bottom-living life from a depth of 33,341 ft (10,162 m) proving beyond a doubt that life exists in the deepest depths of the ocean.

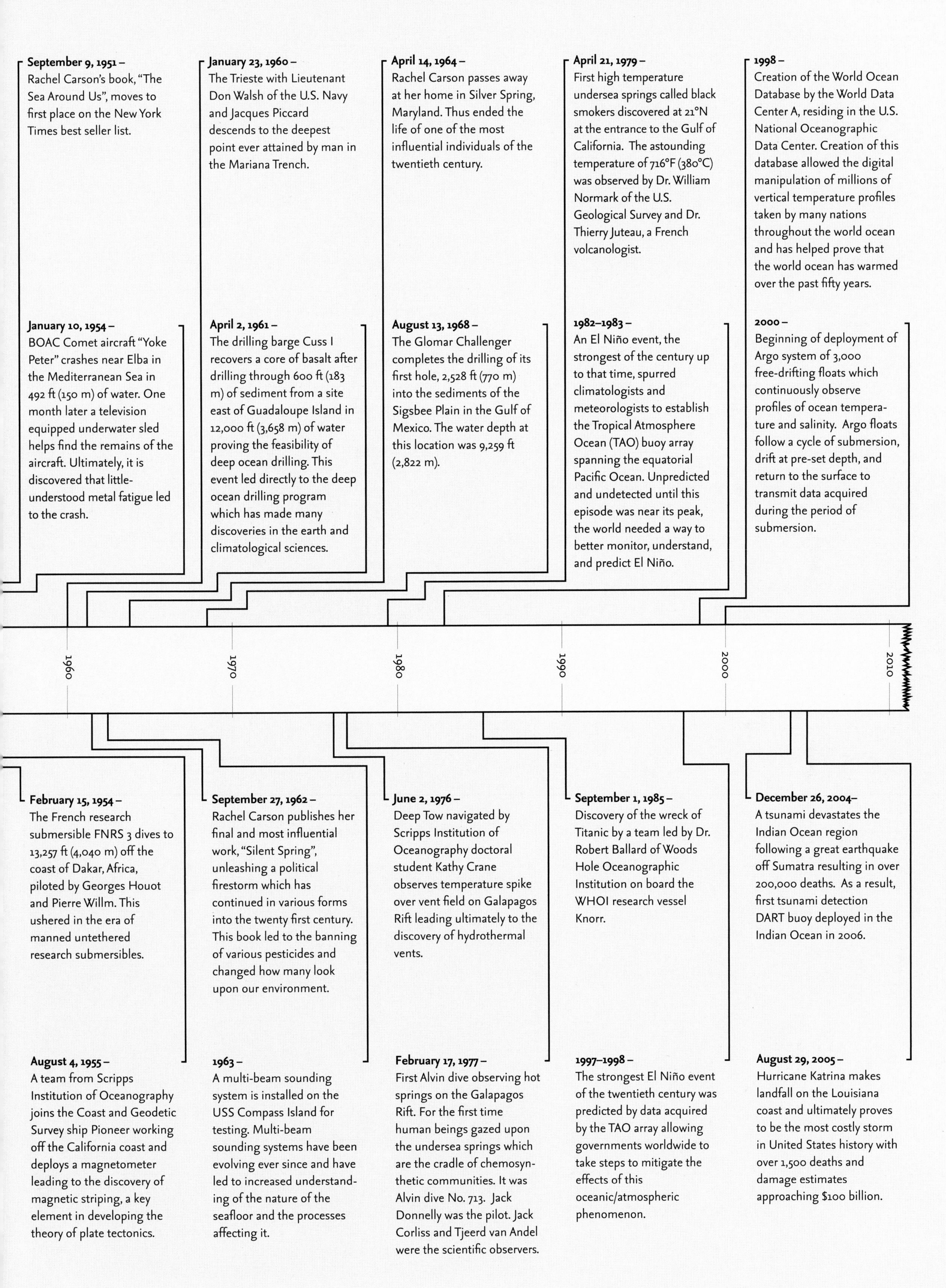
September 9, 1951 – Rachel Carson's book, "The Sea Around Us", moves to first place on the New York Times best seller list.
January 10, 1954 – BOAC Comet aircraft "Yoke Peter" crashes near Elba in the Mediterranean Sea in 492 ft (150 m) of water. One month later a television equipped underwater sled helps find the remains of the aircraft. Ultimately, it is discovered that little-understood metal fatigue led to the crash.
February 15, 1954 – The French research submersible FNRS 3 dives to 13,257 ft (4,040 m) off the coast of Dakar, Africa, piloted by Georges Houot and Pierre Willm. This ushered in the era of manned untethered research submersibles.
August 4, 1955 – A team from Scripps Institution of Oceanography joins the Coast and Geodetic Survey ship Pioneer working off the California coast and deploys a magnetometer leading to the discovery of magnetic striping, a key element in developing the theory of plate tectonics.
January 23, 1960 – The Trieste with Lieutenant Don Walsh of the U.S. Navy and Jacques Piccard descends to the deepest point ever attained by man in the Mariana Trench.
April 2, 1961 – The drilling barge Cuss I recovers a core of basalt after drilling through 600 ft (183 m) of sediment from a site east of Guadaloupe Island in 12,000 ft (3,658 m) of water proving the feasibility of deep ocean drilling. This event led directly to the deep ocean drilling program which has made many discoveries in the earth and climatological sciences.
September 27, 1962 – Rachel Carson publishes her final and most influential work, "Silent Spring", unleashing a political firestorm which has continued in various forms into the twenty first century. This book led to the banning of various pesticides and changed how many look upon our environment.
1963 – A multi-beam sounding system is installed on the USS Compass Island for testing. Multi-beam sounding systems have been evolving ever since and have led to increased understanding of the nature of the seafloor and the processes affecting it.
April 14, 1964 – Rachel Carson passes away at her home in Silver Spring, Maryland. Thus ended the life of one of the most influential individuals of the twentieth century.
August 13, 1968 – The Glomar Challenger completes the drilling of its first hole, 2,528 ft (770 m) into the sediments of the Sigsbee Plain in the Gulf of Mexico. The water depth at this location was 9,259 ft (2,822 m).
June 2, 1976 – Deep Tow navigated by Scripps Institution of Oceanography doctoral student Kathy Crane observes temperature spike over vent field on Galapagos Rift leading ultimately to the discovery of hydrothermal vents.
February 17, 1977 – First Alvin dive observing hot springs on the Galapagos Rift. For the first time human beings gazed upon the undersea springs which are the cradle of chemosynthetic communities. It was Alvin dive No. 713. Jack Donnelly was the pilot. Jack Corliss and Tjeerd van Andel were the scientific observers.
April 21, 1979 – First high temperature undersea springs called black smokers discovered at 21°N at the entrance to the Gulf of California. The astounding temperature of 716°F (380°C) was observed by Dr. William Normark of the U.S. Geological Survey and Dr. Thierry Juteau, a French volcanologist.
1982–1983 – An El Niño event, the strongest of the century up to that time, spurred climatologists and meteorologists to establish the Tropical Atmosphere Ocean (TAO) buoy array spanning the equatorial Pacific Ocean. Unpredicted and undetected until this episode was near its peak, the world needed a way to better monitor, understand, and predict El Niño.
September 1, 1985 – Discovery of the wreck of Titanic by a team led by Dr. Robert Ballard of Woods Hole Oceanographic Institution on board the WHOI research vessel Knorr.
1997–1998 – The strongest El Niño event of the twentieth century was predicted by data acquired by the TAO array allowing governments worldwide to take steps to mitigate the effects of this oceanic/atmospheric phenomenon.
1998 – Creation of the World Ocean Database by the World Data Center A, residing in the U.S. National Oceanographic Data Center. Creation of this database allowed the digital manipulation of millions of vertical temperature profiles taken by many nations throughout the world ocean and has helped prove that the world ocean has warmed over the past fifty years.
2000 – Beginning of deployment of Argo system of 3,000 free-drifting floats which continuously observe profiles of ocean temperature and salinity. Argo floats follow a cycle of submersion, drift at pre-set depth, and return to the surface to transmit data acquired during the period of submersion.
December 26, 2004– A tsunami devastates the Indian Ocean region following a great earthquake off Sumatra resulting in over 200,000 deaths. As a result, first tsunami detection DART buoy deployed in the Indian Ocean in 2006.
August 29, 2005 – Hurricane Katrina makes landfall on the Louisiana coast and ultimately proves to be the most costly storm in United States history with over 1,500 deaths and damage estimates approaching $100 billion.
1960
1970
1980
1990
2000
2010

Exploration technology

Depth of water and location have always concerned the mariner. Accordingly, pole sounding and line and weight sounding were among the earliest of ocean exploration technologies. Navigation was by visual inshore piloting or latitude sailing until the invention in the eighteenth century of the marine chronometer which allowed determination of longitude. In the nineteenth century systems evolved which could measure the greatest ocean depths with some degree of confidence. Hemp rope soundings were first used but the invention of piano-wire sounding in 1872 improved sounding reliability, accuracy, and efficiency. However, fewer than 20,000 deep-ocean soundings had been made by the 1920s when acoustic sounding systems came into general use. Single-beam acoustic sounding systems provided a profile of the seafloor, but the development of multi-beam systems in the 1960s revolutionized our ability to visualize the seafloor by producing swaths of soundings which generate high-resolution and very detailed maps. Navigation systems improved as well with the development of electronic navigation methods during World War II which have culminated in the Global Positioning System (GPS).

Tools for observing biology, geology, chemistry, and physical characteristics of the ocean evolved with sounding systems. Dredges and nets were borrowed from fishermen and modified for oceanography. Pressure-resistant instruments for measuring temperature and obtaining water samples were developed. In 1934 William Beebe pioneered direct deep-sea observation when he sealed himself in a steel sphere with viewing ports and had himself lowered half a mile (800 m) into the sea. The first deep-sea camera was deployed in 1939; side-looking sonars were developed after World War II; and by the 1960s both instruments were intrinsic components of deeply-towed instrument packages. Today, human-occupied vehicles (HOVs), remotely-operated vehicles (ROVs), and autonomous underwater vehicles (AUVs) are all part of the deep-sea explorer's suite of instruments for unraveling the mysteries of the sea.

LEFT:
Marti's continuous recording sounder built by the French engineer Pierre Marti. In 1919, Marti began designing and describing sounding machines based on acoustic methods. This recording device allowed measuring time of sound emanation and time of reception, thus giving travel time which can be used to determine depth.

LEFT:
Johnson Sea Link (JSL) submersible off the coast of Florida. Its 6 inch (152 mm) thick acrylic sphere holds a pilot and an observer.

BELOW:
Life on the Edge 2005 Expedition. The JSL submersible offers a panoramic view of the underwater world . This is a view of dense coral development off North Carolina in about 1,310 ft (400 m) of water.

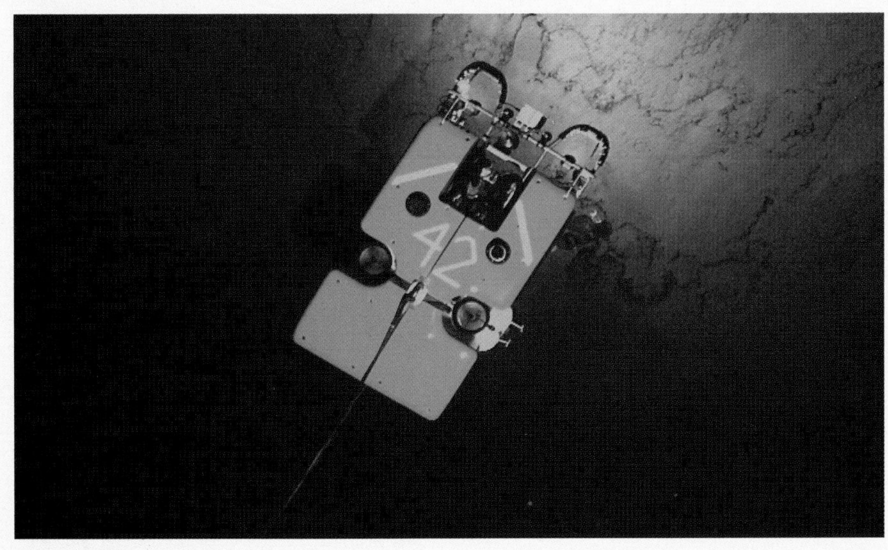

ABOVE:
The remoteley operated vehicle (ROV) Hercules searches for deep sea fauna in the New England Seamount Chain.

TOP:
Hidden Ocean 2005 Expedition. The ROV is brought back on board after a dive into the Canada Basin, Beaufort Sea, Alaska.

ABOVE:
Ring of Fire 2002 Expedition. The crew of the University of Washington research vessel Thomas G. Thompson launching ABE (Automated Benthic Explorer) over the side of the ship for another night of data collection. ABE is outfitted with a host of scientific sensors to log magnetics, temperature, conductivity, and multi-beam bathymetry.

Human interaction

Human interaction
Introduction

The sea has always captured the human imagination in both a mystical and practical sense of discovery and understanding. Sea monsters and tempests threatened the earliest mariners yet sailors still went to sea and landlubbers enjoyed the tales of their exploits. From Homer to Cousteau, human curiosity has been satisfied with the tales and truths of the sea, and motivated the further development of our knowledge and exploitation of this valuable and life giving resource—the living ocean.

Human interaction with the oceans is extensive. Life on our planet has likely come from the sea, evolutionarily, and currently from the oxygen generated by the oceans. Throughout history, the sea has provided lanes of commerce, security, and food, and the military means of competing for these. Science has progressed from suggesting answers to observed or imagined phenomena, to our current understanding of ocean science. Over time this has progressed from the elemental to the complex. Prior to the discovery of hydrothermal vents and the biological communities surrounding them just three decades ago, such questions as how the sea became salty, and how life began on earth seemed reasonably settled. These new discoveries add complexity to our understanding, for example, that salt may more properly be defined to enter the sea from water circulation within the mineral-rich crust rather than from great rains and glacial runoff, and that chemosynthesis may have preceded photosynthesis. Fortunately, what may be the greatest human attribute, reasoned curiosity, will compel further discovery and understanding.

Observing the ocean has generated knowledge of the winds, waves, and depths, from which knowledge of the parameters of safely exploiting the sea has given rise to greater human conveniences such as energy, food, and the military means to protect societies. Our study of marine biology has led to realizations about the diversity of life in the ocean, and the manner in which seemingly alien life forms may hold the answers to our most fundamental questions, such as where life began on earth, and further cures for diseases. By observing the oceans in the most basic context, we have enabled safe and efficient marine transportation by producing maps and charts of ocean approaches to commercial harbors. But while the shallows are mapped, approximately 95 per cent of the world's oceans remain unmapped at any reasonable resolution.

Exploitation and transportation of oil from the sea, and over the sea, has enabled our utilization of fossil fuel in an affordable way. This industry, which brings everyday products and conveniences to us daily, utilizes some of the most complex engineering imaginable, such that the construction and production of deepwater oil platforms are analogous to a lunar landing—truly space age technology applied to the interior of our earth.

On rare occasions, through chance or human error, accidents happen which claim ships and soil the marine environment. Such collisions and spills impact the marine environment to varying extents depending upon the cargo carried, but always invite concern for the mariners aboard, and the downstream or

PREVIOUS PAGE:
Great Barrier Reef, Australia

downwind environment. Utilization of our ocean environment carries a risk and a cost which we have historically chosen to bear, and will likely continue to do so. Shipwrecks from the times of the ancients to as recent as fifty years ago may be described as "historic" and worthy of measures of protection for the value of the historical record they contain. A lively debate continues on the propriety of recovering undersea artifacts for public display or commercial sale, versus leaving these materials in the sea and bringing high definition images of these amazing sites to the surface for free public distribution. In either case, underwater archeology and maritime history reflect human interaction with the oceans through an amazingly rich record revealed by ocean survey and shipwreck mapping.

The oceans hold as much as 98 per cent of the inhabitable biosphere, have given rise to amazingly productive fisheries. During the 1960s and 1970s, most nations looked to the ocean as a great source of protein, a bounty soon harvested for both hunger and novelty. Sustainable fisheries are certainly achievable, and in many cases proper fisheries management has proven so and furthered our human appetite for delicious treats and the marvelous productivity of the oceans. But, in many contrary instances the world's human population has elected to exploit fisheries in excess of a sustainable level, and overfished once-great stocks. Only thirty years ago, what was described as a nearly limitless resource is now found to be quantifiably limited. So too, has the ocean's ability to absorb pollution and byproducts from our civilization, from oil and chemical residues washing from our streets into the rivers, bays, and oceans, to the great amounts of carbon compounds from combustion, being absorbed by the ocean. The seeming endless ability of the oceans to absorb, dilute, and buffer the impact of human interaction has become apparent and palpable. As carbon loading of the oceans is found to contribute to ocean acidification, we can find plastic residues of our civilization—cigarette lighters and soap containers—at the surface and in the deepest parts of the ocean, at the greatest distances from land.

Many of the public choices confronting the nations of the world are dependent upon the quality of the information available to decision makers. This quest for knowledge drives the need for continued ocean science and monitoring, and the need for public education to establish ocean literacy. Perhaps the single greatest advance in public interest and understanding of the oceans was the development and production of SCUBA equipment, enabling millions of the earth's citizens to personally experience the wonders of the ocean world. As multiple generations have marveled at the images of coral reefs, either on television or in person, an awareness and concern for ocean life has increased, bringing about laws relating to clean water and marine area protection. Soon, with the increased availability of deepwater camera systems and live satellite transmissions, the equally colorful images of deep sea communities will inspire the next generations of explorers. Human interaction with the oceans is inextinguishable.

NOAA TAO

Ocean observation

Having information about our environment is key to sustaining the earth, our way of life, and future generations. In 1990, twenty three per cent of the world's population (or 1.2 billion people) lived both within 60 miles (100 km) distance and 300 ft (100 m) elevation of the coast, at densities about three times higher than the global average. By 2010, twenty out of thirty mega-cities will be along the coast with many low-lying locations threatened by sea-level rise. Coastal storms account for over 70 per cent of recent U.S. disaster losses annually. Twenty-five per cent of earth's biological productivity and an estimated 80-90 per cent of global commercial fish catch is concentrated in coastal zones. Enhanced observations and data integration will improve our early warning and forecasting abilities, allowing our coastal communities to better prepare and respond to potential danger. Integrating observations and services will expand information for people, communities, states, nations, and global populations as pubic health, global commerce, and environmental conditions are better understood. Greater access to data and services is a significant contribution to future accomplishments.

So why is observing the ocean so important to the U.S.? Due to population growth along U.S. coasts, as well as their strong connection to the national economy through maritime commerce and tourism, and the environmental and climatological importance of coastal and ocean ecosystems, the physical and ecological states of these systems have disproportional effects on the safety and wellbeing of human populations (Integrated Global Observing Strategy (IGOS) Coastal Theme Report, 2006). The indirect impacts of disruptions in these areas due to natural hazards, for example, can be far-reaching if infrastructure and services such as ports, oil refineries, waste-water treatment facilities, agriculture, and power plants are disrupted or disabled.

This coordinated data network will allow the scientist, farmer, teacher, emergency responder, environmental resource manager, and many others rapid access to comprehensive information on demand, and in formats which are useful for making everyday decisions and improving our overall quality of life.

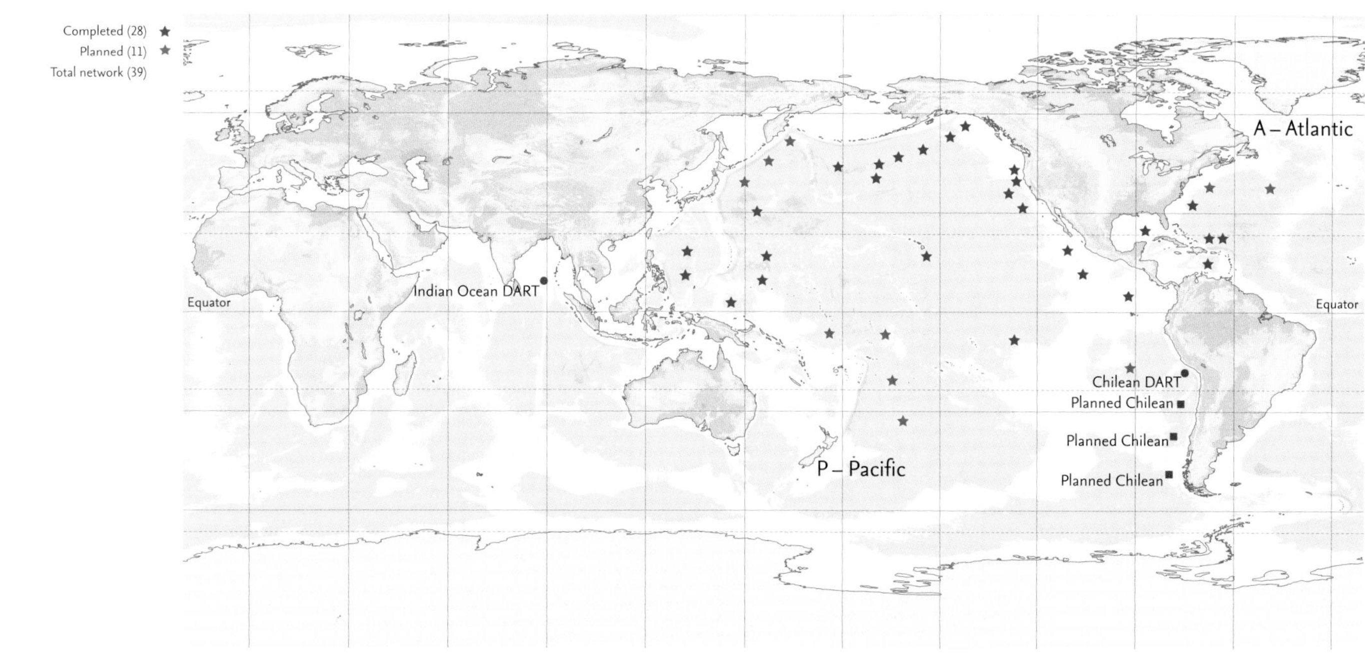

TOP LEFT:
One of fifty-five buoys serviced by the National Data Buoy Center (NDBC) along the equator in the Pacific Ocean.

LEFT:
European Remote Sensing Satellite for measuring environmental factors such as sea surface temperature.

ABOVE:
Worldwide distribution of Deep Ocean Assessment and Reporting of Tsunamis (DART) buoys. The DART program is an ongoing effort to maintain and improve the capability for the early detection and real-time reporting of tsunamis in the open ocean.

Diving

Diving is the most intimate human interaction with the world's oceans. A diver can see, hear, and feel the undersea environment while the brain is recording images and impressions in context with the surroundings. Divers may dive for recreation, or for work, but all leave the water with a greater appreciation of the ocean. A scientific diver enters the water with the primary purpose of adding to our knowledge of what the ocean is, and how we can both use it and better understand it while managing and protecting the living and non-living resources it encompasses.

The use of a self contained underwater breathing apparatus (SCUBA) allows free swimming scientists to make human observations and conduct fine-scale manipulations impossible by any other means. Most SCUBA dives are shallower than 130 feet (40 m) and use air as the breathing gas. The technology now exists for SCUBA divers to extend their working depth in excess of 300 feet (91 m) by breathing a mixture of oxygen, nitrogen, and helium in open circuit mode or using closed-circuit mixed-gas rebreathers.

The National Oceanic and Atmospheric Administration (NOAA) Dive Program trains divers and diving physicians in the latest diving technologies, and combines with the Undersea Research Program to conduct studies to increase the safety and efficiency of divers. NOAA sponsors 25,000 dives per year in support of fisheries management, marine protected areas, ship operations, and containment of hazardous materials. United States agencies sponsor an estimated 120,000 world-wide scientific dives annually.

TOP:
A NOAA diver, utilizing untethered, mixed-gas diving techniques, conducts research at the Monitor National Marine Sanctuary in support of major archaeological and engineering recovery expeditions conducted by NOAA and the U.S. Navy during 1998-2002.

BOTTOM:
An archaeologist "flies" a photo-mosaic sled over the shipwreck Benwood in the Florida Keys National Marine Sanctuary. The National Marine Sanctuaries Maritime Heritage Program has conducted mosaic surveys in several sanctuaries including Monitor, Thunder Bay, and the Florida Keys.

Submersible technology

Since humans first gazed upon the oceans they have wondered "what mysteries lie beneath the surface." Early observations from sailors, fishermen, and others who relied upon the oceans for commerce demonstrated that the oceans were full of strange creatures, unpredictable currents, even deep canyons, and extensive mountain ranges. However, unlike astronomers who could observe their subjects directly and more accurately with increasingly powerful instruments, those interested in exploring the ocean depths were challenged by the seemingly impenetrable waters themselves.

As early as the mid-1500s inventors designed and experimented with submersibles as a means to gain access to and observe this unknown submerged world. However, it was not until the advent of the Industrial Revolution that the pace of these ventures accelerated and the number of new and more effective systems increased. Today, submersibles can be classified in three broadly defined categories: human occupied vehicles (HOV); remotely operated vehicles (ROV); and autonomous underwater vehicles (AUV). Each has its own unique capabilities to further advance efforts to discover new ecosystems and species, and to learn more about the role the oceans play in sustaining life on the planet.

With 95 per cent of the world's oceans unknown and unexplored, we can continue to expect rapid changes in these systems, as well as in the sensors that they carry. We can expect changes in submersible technology to help us answer existing questions related to ocean resources and environmental challenges, as well as to provide future generations of scientists and explorers continued and improved access to what is truly the earth's final frontier.

TOP:
The Hercules ROV exploring a seamount off the coast of New England, U.S.A.. The Hercules is owned and operated by the Institute for Exploration (IFE) and works in tandem with a mobile camera sled named Argus.

BOTTOM:
The Johnson Sea-Link (JSL) is a four-person HOV owned and operated by the Harbor Branch Oceanographic Institution. Here, the JSL is recovering a specialized bait trap and ambient light camera which captures images of fish and other species which typically avoid the high-intensity lights mounted on submersibles.

Marine biology

As you slip under the surface of the water 100 miles (161 km) south of the Texas/Louisiana border in the Gulf of Mexico, at a little known dive destination called the Flower Garden Banks National Marine Sanctuary, you'll be greeted by as much of one hundred feet (30 m) visibility, and coral for as far as you can see! You may encounter large loggerhead sea turtles resting under coral ledges on the reef, or graceful manta rays which glide overhead, and as you watch, might unfurl their cephalic fins – horn-like fins which protrude from their heads, open their mouths wide, and with a flip of their wings, roll into a slow barrel roll, an action whereby they filter water and plankton through their gills. The reefs themselves are dominated by enormous heads of star and brain coral, which provide habitat for hundreds of species of crabs, lobsters, snails, sponges, fishes, and algae. Stunning natural events occur each year here – schooling hammerhead sharks and spotted eagle rays, and mass spawning of corals and sponges!

The coral reefs of the Flower Garden Banks are but a small portion of the areas contained within the sanctuary boundaries – the deepwater portions, the majority of the sanctuary, and less well known, but just as spectacular, and as valuable to the ecosystem as the coral reef. These deepwater areas are home to vibrantly colored sea fans and sea whips – reds, yellows, oranges! Deepwater fishes, which no diver will ever see, school in their thousands, and forage and reproduce in forests of black coral and sponges. Deepwater technologies have allowed us to view spectacular eruptions of underwater mud volcanoes, and watch ripples on the surface of super-saline underwater rivers flowing through the ocean.

These reefs are three of dozens of reefs and banks which parallel the continental shelf edge. We are learning more about the deeper areas of these areas through technology which allows us to stay deeper and longer than ever before. We are guided through these areas by high resolution bathymetric maps which we can use just like road maps, so we know where we are, where we've been, and where we will go next. Exploration will help to uncover the intricate biological, ecological, and geological webs which no doubt bind together this mosaic of reefs and banks of the northwestern Gulf of Mexico.

ABOVE:
Dozens of reefs and banks harboring rich biological communities parallel the edge of the continental shelf in the northwestern Gulf of Mexico. Several sites which have recently been explored by the Flower Garden Banks research team are labeled in this regional chart.

TOP:
A ruby brittle star takes advantage of an easy meal as a star coral releases bundles of gametes in a highly synchronized coral spawning event at the Flower Garden Banks.

TOP RIGHT:
A diver encounters a large loggerhead sea turtle resting in a sand flat at the East Flower Garden Bank.

RIGHT:
A school of creole-fish on the McGrail Bank, Gulf of Mexico.

FAR RIGHT:
A searobin. When swimming, its large pectoral fins open and close like a bird's wings. The fish can use its pelvic fins to "walk" along the sea floor.

Nature and industry sharing common ground

Imagine yourself on a ship, one hundred miles away from land, with nothing but deep blue water surrounding you. Oh, and several dozen hulking structures that are oil and gas platforms and drilling rigs. Welcome to the Gulf of Mexico.

We live in a society which relies heavily on the oil and gas industry to function. The Gulf of Mexico is one of the most active oil and gas production fields in the world, with close to 4,000 platforms extracting the precious resources and delivering them to shore via pipelines and ships.

What may come as a surprise to many, is that as close as one mile from one of these gas platforms, is one of the most pristine coral reef systems in the Caribbean – the magnificent reefs of the Flower Garden Banks. There are many reef systems worldwide which share space with industry. Examples include: Ningaloo Reef in Australia, The Gulf reefs in the Middle East, and recently, reefs of Belize.

Communication is critical to the coexistence of marine resources and industry. Environmental stewards and industry managers should strive to accept and acknowledge each user group's role and discuss the activities openly, and develop plans to conduct activities which are environmentally sound. The groups should be prepared to invest heavily in scientifically-based monitoring programs which will be able to detect small changes in the health of the potentially affected ecosystems, and report in a timely manner. Response actions should be set forth up front, and agreed upon by all parties.

The reefs and banks in the northwestern Gulf of Mexico have coexisted with oil and gas production for the past forty years. Minerals Management Service (MMS), a part of the U.S. Department of Interior, regulates the Gulf of Mexico oil and gas industry to protect the biological environments which share the sea floor with industry. NOAA's National Marine Sanctuary Program (NMSP), in turn, is also charged with the management and protection of the specific natural resources, and works closely with MMS to conduct monitoring and research necessary for the protection of the resources, as well as developing strategies for protection, not only from oil and gas, but from other uses such as fishing, diving, and extraction of other kinds.

RIGHT:
Oil and gas platforms in the Gulf of Mexico.

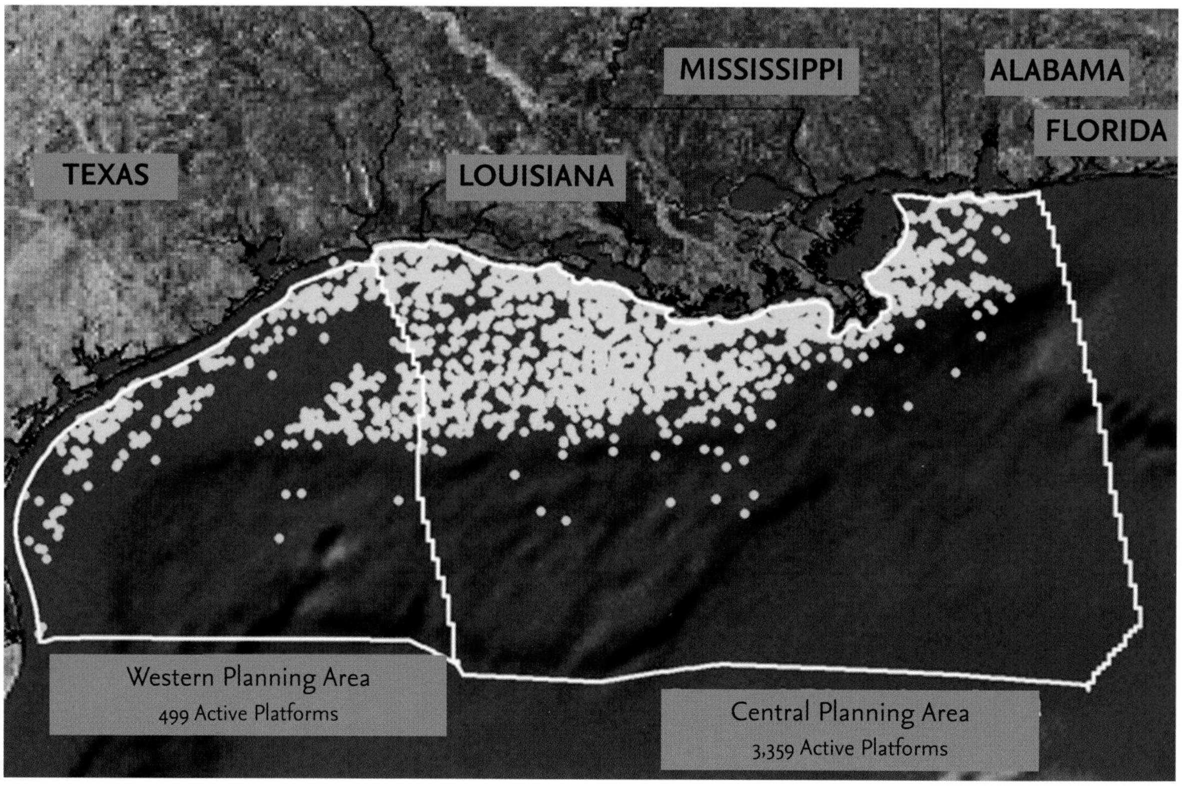

TOP RIGHT:
An offshore oil production rig off the coast of Texas. The vast majority of U.S. offshore rigs are located in this area.

RIGHT:
A living cloud of fish in the Flower Garden Banks National Marine Sanctuary, Gulf of Mexico.

LEFT:
Container ships passing in Elliot Bay, Washington, U.S.A..

Commerce

The oceans and seas are the most efficient and economical means of transporting goods and people from place to place. Essential for global commerce, they connect landlocked roads and railways to the rest of the world. A country's ability to maintain safe and efficient navigation routes within its territorial waters and to manage its port infrastructure ensures that waterborne commerce and transportation continue to flourish.

Phoenicia was the earliest recorded civilization to conduct shipbuilding and expansive trade across the Mediterranean Sea. By means of man-powered sailing vessels, the Phoenicians set up numerous commercial outposts and traded items such as cedar, glass, and textiles for tin, silver, and copper with their Mediterranean neighbors. As a result, this entrepreneurial society was extremely profitable from 1200 to 800 BC. Ship-based transits between peoples and countries spanned the known world, fostering exploration and the exchange of goods, cultures, and ideas, along with disease and wars in the quest for wealth and power.

Jumping ahead many centuries, the Industrial Revolution in the eighteenth and early nineteenth century marked another huge leap in maritime commerce with the development of the steam-powered ship. The efficiency of steam power, and later diesel fuel, meant that ships could go further with more cargo and in less time. In the 1950s, cargo shipping again experienced a radical change with the adoption of metal containers which could be craned directly between trucks or trains and the ship. The resulting growth in container shipping has led to dramatic increases in ship size and numbers. To respond, ports and mariners are utilizing tools like electronic nautical charts, real time water levels and high-accuracy satellite positioning capability to bring ships in and out with as much cargo as they can carry safely. This pace seems unlikely to slow down; by 2020, international maritime trade is expected to double, or even triple. As of 2006, waterborne commerce remains the backbone of the U.S. economy, contributing over 13 million jobs and $1 trillion annually to the U.S. Gross Domestic Product.

ABOVE:
Emma Mærsk, the largest container ship in the world. Emma Mærsk is able to carry 1,400 more containers than any other ship is capable of carrying.

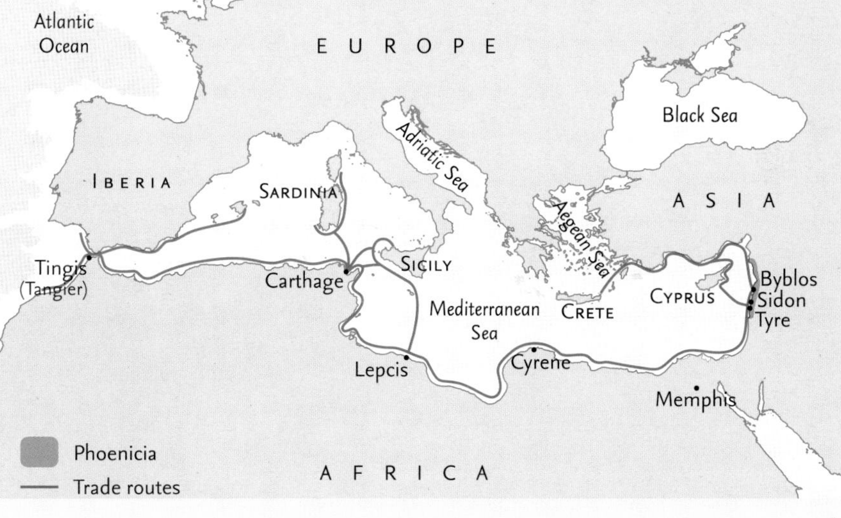

RIGHT:
Commercial network of the Phoenicians.

Seaports

Once mankind launched vessels capable of plying the oceans and carrying goods and passengers, seaports came into being. Seaports were and still are generally located where the coastline provides shelter from the elements, sufficient water depth, and convenient access to land transportation routes. Seaports also serve as coastal defenses from which vessels of war can sally forth to defend a nation's shores.

Seaports draw people to live and work near the coast. They connect countries to each other through shipping routes, provide a nexus for cultures to mix, and spawn great cities. For example, Ostia Antica, founded around the fourth century BC, is well known as ancient Rome's seaport on the Tyrrhenian Sea. Ostia protected Rome from seaborne invasions and enabled cargo vessels from countries around the Mediterranean and beyond to trade with Rome. By the second century AD, it had become a flourishing commercial center inhabited by upwards of 100,000 people.

Today seaports still serve as gateways to our countries. They are vital to our commerce, our national security, and ultimately our economic health. Cargo, commodities, people, and energy supplies, almost everything imaginable, transit our seaports to or from intermodal transportation nodes. A farmer in Kansas may well be exporting his crop overseas to Europe, or wearing clothes imported from Asia, both of which pass through seaports. The modern seaport must keep pace with the massive volume of material which transits through it on a daily basis to satisfy a rapidly growing world population, a global marketplace unfettered by geography, just-in-time business practices, and ever-larger, deeper, and taller vessels exchanging goods between cultures and countries.

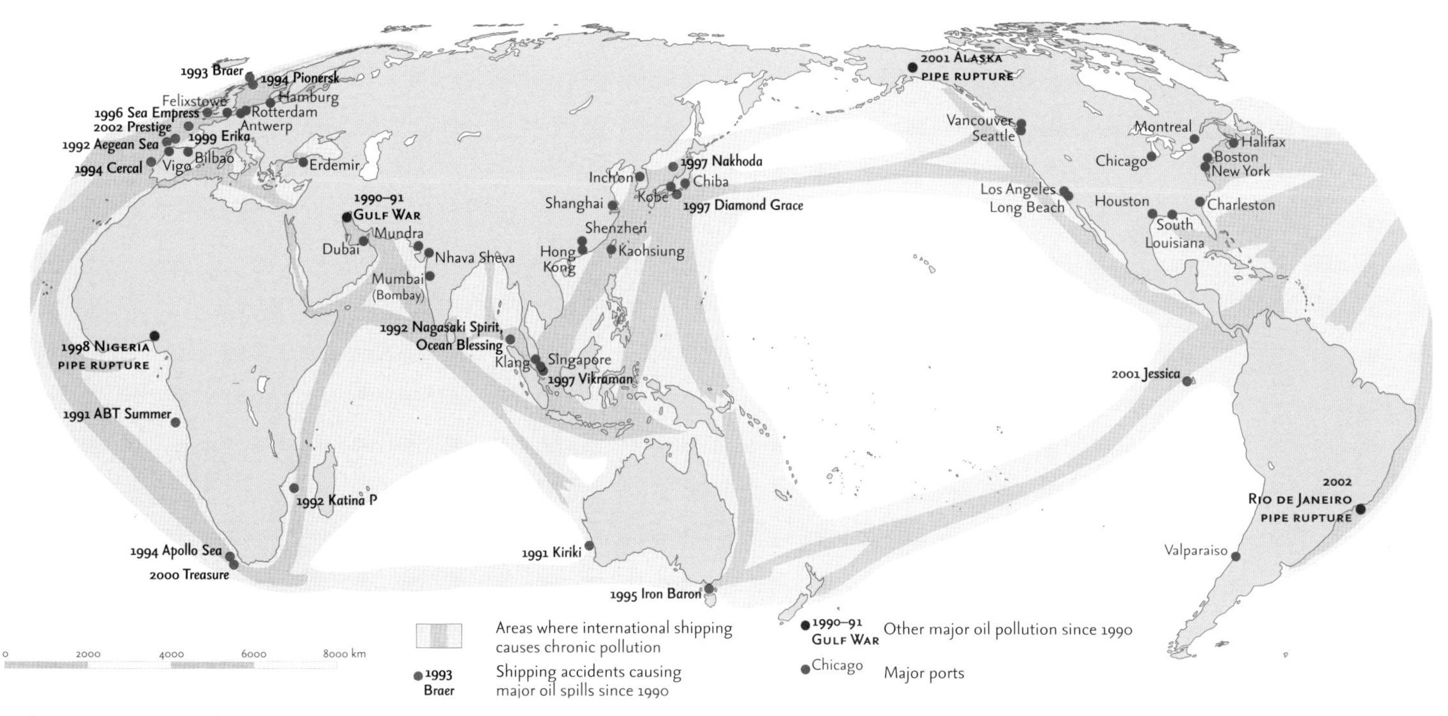

Shipping routes

ABOVE:
The Esteem Splendour making her approach at the Houston/Galveston Seabuoy. To get an idea of the size of this ship, there is a person at the head and the foot of the gangway.

LEFT:
The world's major ports and pollution problems caused by international shipping.

The flourishing spice trade between Europe and India led Vasco da Gama to round Cape Horn in 1497 in search of a quicker route, but it took until 1869 for humanity to improve upon his shortcut with the completion of the Suez Canal. Throughout history, commerce has driven the quest to find faster, safer and more economical means of transportation, and the solution has been consistent—the ocean. However, the story of waterborne commerce has been anything but trouble-free. Geographic chokepoints like the Straits of Hormuz and Gibraltar have seen numerous international confrontations. Man has also improved upon nature to increase efficiency in trade by creating waterways like the Suez and Panama Canals.

Today commerce, engineering, geography, and the environment meet in the time-sensitive high-traffic setting that is a modern port. With nearly $1 trillion worth of cargo per year moving through the west coast ports of North America alone, an interruption at a major port is quickly felt throughout the country. For example, a strike at the port of Los Angeles in 2004 caused an average delay of seven days to unload the ships calling on the port. This affected imports from Asia all across the U.S., with impacts such as food rotting dockside and toys arriving late for the holidays. After Hurricane Katrina in 2005, agricultural commerce was adversely affected in America's interior because ocean-going ships had to wait at the mouth of the Mississippi River for Gulf port facilities to reopen.

The tools for managing these port areas, including nautical charts, tide data, and accurate geographic positioning, help to make possible an uninterrupted flow of traffic. If the physical dimensions of a port or channel are not properly measured and displayed for the mariner, then the facilities themselves are not safe. The increasing size of ships places a strain on the infrastructure which serves them, making the measurement of every inch of depth, width, and height critical to safe navigation.

Predicting shipwreck locations

FAR RIGHT:
Scuba diver near wreck off the coast of Cuba.

BELOW:
3-D rendering of a very high-resolution multi-beam survey of Mulberry Harbor off Omaha Beach, the famous World War II invasion site. Several sunken caissons and a blockship which had been deployed to add substance to the artificial harbor can be seen in this image as well as sedimentary bedforms which clearly indicate the direction of current flow in the region and the effect of that flow on the wreckage.

A confluence of several indicators was known to create traps for sailing ships in U.S. waters between the sixteenth and early twentieth centuries. Studying those factors and applying them to coastal geography allows marine archaeologists to predict areas of likely location for historic shipwrecks.

Archaeologists recognize that besides port location, indices which threatened ships included: location of shipping routes; natural submerged and shoreline landscape features such as shoals, reefs, sandbars, and barrier islands; ocean currents and winds; and historic hurricane routes.

The greater the number of these factors which were present in an area, the higher the likelihood of shipwrecks. Ships naturally must come to a port. The larger and more economically vibrant the port, the more watercraft were drawn to it. In addition to ocean-going ships, a port contained numerous smaller vessels associated with local trade and port business as well (messenger vessels, yard boats, tenders, pilots, etc.). The greater the number of vessels which traversed the waters, the higher the odds became that craft might have accidents.

Add fickle weather to crowded navigation lanes and the odds of shipwreck became decidedly worse.

Unforgiving natural landscape conspired to wreck a ship by providing checks to vessels' ability to maneuver which were not present on open water. As if in a deadly chess game, the environment and other players could force a ship aground.

A quick glance at the sonar image of the wrecks of the schooners Louise B. Crary and Frank A. Palmer (wrecked 1902) provide evidence of merging shipwreck indices. Now resting on the Stellwagen Bank National Marine Sanctuary's seabed, the two ships collided at the mouth of Massachusetts Bay, vying to be the first into Boston. Their wreck was made likely by a confluence of shipping route, proximity to port and the need to stay clear of Cape Ann's rocks.

In Lake Huron's Thunder Bay National Marine Sanctuary, archaeologists study the remains of the schooner E.B. Allen, lost in 1871 after colliding with another ship in fog near land. The indices contributing to the Allen's loss include a heavily trafficked shipping lane near the coast, and the complications brought by weather.

Mechanical power, navigational aids and radar have drastically reduced the threat of the shipwreck indices. However, these factors are always in place waiting for new incautious sailors.

BELOW:
3-D rendering of a sunken shipwreck.

BOTTOM:
Side scan sonar image of the Palmer and Crary wrecksite on the seabed of the Stellwagen Bank National Marine Sanctuary. The Louise B. Crary is the southern (bottom) ship and the Frank A. Palmer lies to the north (top of picture).

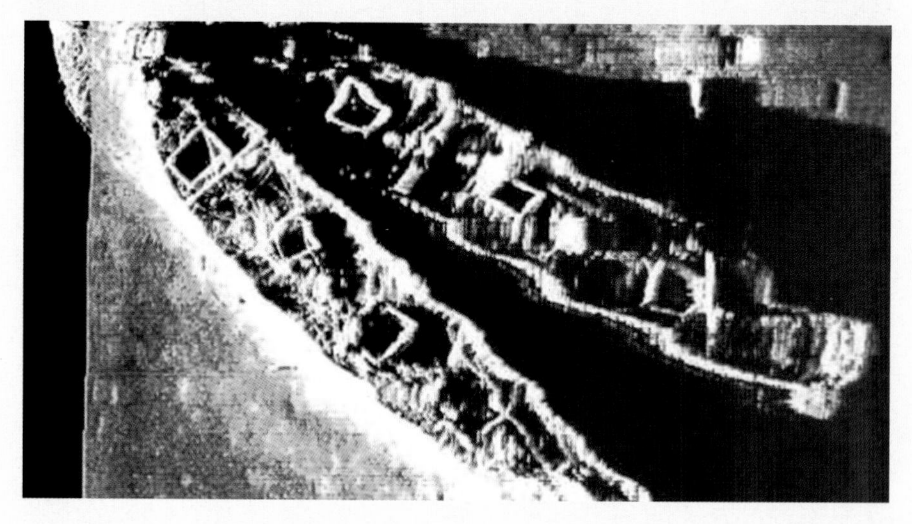

Ancient shipwrecks of the Mediterranean and Black Seas

The deep sea remains a mysterious place on planet earth, but it is beginning to reveal important details about the natural and human history of our planet. Shipwrecks, especially well-preserved shipwrecks in the deep sea, are veritable time capsules which can reveal fascinating elements of our past. This is especially true in the Mediterranean and Black Seas, where human history spans many thousands of years, from Paleolithic times to the present day. This region also represents a cultural crossroads linking the populations of ancient Europe, Asia, Africa, and the Middle East. Shipwrecks discovered in this region, which range in age from the Late Bronze Age more than 3,000 years ago to the present day, have revealed new information which has helped to rewrite text books on ancient history.

One very significant shipwreck which was discovered off southern Turkey at Uluburun took more than ten years to excavate by divers and archaeologists affiliated with the Institute of Nautical Archaeology (INA) located at Texas A&M University. This Late Bronze Age shipwreck carried a wealth of raw materials which were fit for a king, including copper and tin ingots, ceramic jars filled with incense, hand-made bronze tools, beautifully colored glass ingots, ivory, beads, and jewelry. Two other significant shipwrecks from this region were discovered off the coast of Israel in water deeper than 1,312 feet (400 m) and explored using remotely operated vehicle (ROV) systems by a research team from Harvard University and the Institute for Exploration in Mystic, Connecticut. These shipwrecks, Tanit and Elissa, which contained thousands of ceramic jars which carried wine and other commodities, were likely sailed by ancient Phoenicians during the Iron Age who were trading goods between ports in Israel and Egypt around 800 BC. Lastly, one of the best-preserved ancient shipwrecks was discovered in the Black Sea off Sinop, Turkey by a team of researchers affiliated with the Institute for Exploration. This shipwreck, Sinop D, dates to the Byzantine Period nearly 1,500 years ago, and was found within the Black Sea's anoxic layer, where the lack of oxygen prevents the ship's decay by wood-boring organisms. The wooden structural members of the ship's hull and mast were perfectly preserved, as were the small ceramic jars the ship carried as cargo. Through ongoing work by the Institute for Archaeological Oceanography at the University of Rhode Island, this shipwreck will continue to provide new information about the construction of ships in ancient times and details about trade and culture throughout this region.

The seafloor of the Mediterranean and Black Seas represents a nearly unexplored underwater museum which contains thousands of artifacts from antiquity. The shipwrecks found here can tell us a great deal about human interaction with the oceans, through seafaring, ancient trade, and cultural contact. We are just beginning to explore the depths of these seas and we can look forward to seeing their secrets revealed.

RIGHT:
In the Aegean Sea, which is an arm of the Mediterranean Sea, a diver holds a ceramic vessel from the remains of a ship wrecked in 1025 AD.

LEFT:
High definition underwater still images of two ancient shipwrecks: Wooden structural frame of Sinop D (left) from the Black Sea off northern Turkey being brushed by the Hercules ROV, and an amphora from Skerki D (far left and middle) being retrieved by Hercules and brought onboard for conservation and study.

Fishing and fishing grounds

Fisheries worldwide provide valuable sources of food, employment, and revenue for populations in both the developed and developing nations. Globally, commercial fishing had an estimated first-sale value in 2004 of approximately U.S.$85 billion. The value of fisheries to the world economy is far greater. For example, the value of U.S. commercial fish landings of U.S.$4.0 billion in 2006 resulted in an estimated contribution to the U.S. economy of U.S.$35.1 billion. Fishing also provides a significant source of employment. The number of people around the world fishing and fish farming has increased over the last three decades and was approximately 41 million people in 2004.

In 2005 approximately 84 million metric tonnes of marine fishery resources such as fish and invertebrates were caught globally. To get a sense of how large this is, imagine New York City's tallest building, the Empire State Building which is 1454 feet (443m) tall and weighs 365,000 metric tonnes. It would take the weight of over 230 buildings to approximate the world catch of marine fish, crustaceans, and mollusks. Also in 2005, aquaculture, the farming of aquatic organisms, provided approximately 19 million metric tonnes of marine fish, crustaceans, and mollusks. Using the Empire State Building example above, this would amount to over the weight of 52 buildings. In addition to marine animals, marine plants such as seaweed are also harvested for food, biomedical applications, and other purposes.

Some of the predominant fishing nations (ranked by total catch weight) include China, Peru, the U.S., Indonesia, and Chile. Highly productive fishing areas include the northwest Pacific, southeast Pacific, western central Pacific, and northeast Atlantic. These four areas combined produced over 60 % of the 2004 world marine catches. Important species on a global scale include anchovies, herring, sardines, codlike fishes (e.g. cods, hakes, haddocks), and tunas, all of which maybe used either directly for human consumption or for industrial purposes such as meal or oil.

The ocean, however, is not without its limitations. While many fish stocks are managed sustainably, approximately 25 per cent of stocks require further actions to remain sustainable worldwide. Reducing bycatch, marine life caught incidental to the target species, is also a challenge. Nations must continue to work together to help build and maintain sustainable and healthy marine resources important to commercial, recreational, and subsistence fisheries.

RIGHT:
A fishing boat returning from the Adriatic Sea, Rovinj, Croatia.

BELOW RIGHT:
Etang de Thau and oyster farming in Languedoc Roussillon, France.

BELOW:
As well as the intended catch, fishing nets can collect unwanted and even endangered species. Photograph from Pittenweem, Scotland.

Pollution

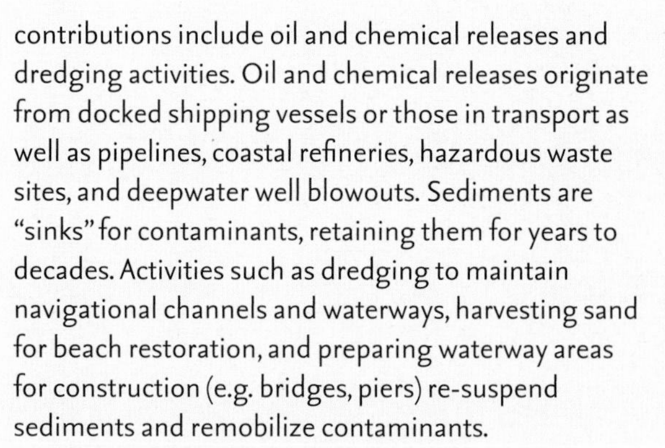

RIGHT:
Container vessel MSC Napoli beached near Sidmouth, England in January, 2007.

BELOW:
An endangered Hawaiian monk seal entangled in a derelict buoy and rope may be unable to dive for food, or swim quickly enough to escape from a predator. Entanglement in marine debris is one of the main causes of mortality in monk seals.

On January 18, 2007 the MSC Napoli, a container vessel en route to Portugal, encountered high seas and gale-force winds which caused significant damage to the ship's hull and flooding in the engine room. After sending out a distress signal, the crew abandoned ship. The ship was later taken under tow in hopes of reaching a designated port some 140 miles (225 km) away. In danger of breaking apart, the ship was beached near Sidmouth, U.K.. While en route to its grounding location, the Napoli lost a portion of its cargo including barrels of wine, motorcycles, exhaust pipes, and diapers as well as hazardous materials such as battery acid, perfume, and the ship's oil supply. The Napoli incident provides one example of human impact on the earth's ocean.

Pollution from numerous sources—point (localized) and non-point (diffuse) —contributes to people's overall impact on the ocean. Point source contributions include oil and chemical releases and dredging activities. Oil and chemical releases originate from docked shipping vessels or those in transport as well as pipelines, coastal refineries, hazardous waste sites, and deepwater well blowouts. Sediments are "sinks" for contaminants, retaining them for years to decades. Activities such as dredging to maintain navigational channels and waterways, harvesting sand for beach restoration, and preparing waterway areas for construction (e.g. bridges, piers) re-suspend sediments and remobilize contaminants.

Non-point source contributors include marine debris (e.g. any anthropogenic object discarded, disposed, or abandoned which enters the coastal or marine environment) and storm water and agricultural runoff. Runoff results from rain or snow melt which transports pollutants to rivers, lakes, and coastal marine environments.

Marine debris

BOTTOM LEFT:
A threatened green sea turtle entangled in a derelict net may drown because it is unable to breathe at the ocean's surface, or starve because it cannot reach food. This sea turtle was fortunate to be released from the net, but many do not survive.

LEFT:
Divers with the National Oceanic and Atmospheric Administration's Fisheries Service use floats to raise a large bundle of derelict fishing gear, which was snagged on a coral reef in the northwestern Hawaiian Islands.

Marine debris has become one of the most pervasive pollution problems facing the world's oceans and waters. As society has developed new uses for plastics, the variety and quantity of synthetic items has increased dramatically, allowing our oceans to be littered with everything from soda cans and plastic bags to derelict fishing gear and abandoned vessels, most of which will likely take decades to centuries to degrade. Marine debris may come directly from ocean-based sources, such as ships and fishing vessels, or indirectly from land-based sources when washed out to sea via waves, rivers, streams, or storm drains. Some debris may be intentionally dumped, but some may be swept or blown into the ocean during storms or lost by accident.

Marine debris not only affects the beauty of our environment, it injures and kills marine life, damages marine habitat, interferes with safe navigation, impacts our shipping and coastal industries, and poses a threat to human health. Marine debris can take many forms, from a speck of plastic to a bundle of fishing nets which can weigh several tons, to an entire ship broken apart and crushing coral reefs in a severe storm. Small pieces of floating plastic are mistakenly eaten by seabirds, which can starve to death with their stomachs full of plastic items. It can entangle sea turtles, whales, and seals, slowing their swimming or trapping them underwater, where they can drown. Derelict fishing gear is a particular menace as nets, lines, and traps can continue "ghost fishing" for years. Marine life caught in the debris dies and then attracts and entangles other creatures. This cycle of death continues until the derelict gear is removed or deteriorates.

Current efforts to address the problem of marine debris are promising; some countries, states, and cities have banned the use of plastic shopping bags, and thousands of people participate in coastal cleanups. Still, more can be done. Individuals can take responsibility for their own litter, industries can decrease their contribution to marine debris, and governments, industries, and nongovernmental organizations can work together to find solutions.

Oil

TOP RIGHT:
Crews containing a leaking oil well in Bayou Perot, Louisiana, U.S.A., in January 2007. The incident occurred when a vessel hit a wellhead. Much of the spilled oil stranded in nearby wetlands, and cleanup lasted for several months.

Catastrophic oil spills are one of the iconic images of technological disasters. Most people readily empathize with the images of oiled seabirds, or the apparent futility of trying to clean up after a large spill.

Thousands of oil spills occur every year in the world's oceans. Oil tankers are commonly thought to be the largest source of spills; however, spills come from a variety of sources, such as production platforms, pipelines, barges, non-tank vessels, and recreational vessels. Many scientists believe that low-level but chronic surface run-off from highways and roads is a major source of marine oil pollution.

Spilled oil threatens marine life and coastal habitats in several ways. Spills can harm plants and animals through physical contact, such as directly coating and smothering them, or by matting their fur or feathers, causing them to die from hypothermia. Spills can also cause toxic contamination, when organisms ingest oil particles, or inhale hazardous vapors. Spills can also cause the destruction of food resources, by tainting organisms which are prey for others.

Oil spill cleanup is generally quite challenging. Most oils float, but a few sink to the bottom. Some are very persistent and sticky, while others disperse and spread rapidly. There are a number of technologies to control and recover spilled oil, but cleanup is still a dirty and difficult process. Prevention of spills is the best solution and new laws in the U.S. and elsewhere have helped to reduce the number of spills.

Energy

Oceans contain enormous amounts of energy which can help the world meet increasing energy demands. By 2030, population growth and economic development are projected to increase energy consumption by 71 per cent from 2003 levels, putting additional pressure on major existing fuel sources. If only 0.2 per cent of the ocean's untapped energy could be harnessed, however, it is estimated that it could provide power for the entire world.

Oil and gas are currently the largest energy sources which are harnessed from the ocean. These fossil fuels are located in large deposits in the seabed. More than a quarter of the oil and gas produced in the U.S. comes from offshore areas, and many other countries have extensive offshore oil and gas facilities as well.

Oil and gas reserves are only a small fraction of the total energy in our ocean. The ocean is also a vast source for renewable energy sources such as offshore wave, tidal, current, wind, salinity gradient, and thermal gradient energy sources. These resources have great potential for meeting increasing energy demands because they are regenerative and cannot be depleted, and do not produce as many greenhouse gases and other pollutants as fossil fuel combustion. Offshore renewable energy technologies, however, are currently used to a much lesser extent for commercial energy generation, and few countries have extensive offshore renewable energy facilities. Increases in energy demand and fossil fuel prices, however, have made these renewable energy sources more economically competitive and some are starting to become utilized. For instance, offshore wind turbines are utilized in numerous European countries to supplement their energy needs. Currently, Denmark generates twenty per cent of the nation's electricity supply from turbines. Additionally, interest in the U.S. is starting to increase and numerous projects have been proposed.

RIGHT:
The North Sea, to the east of the U.K., contains large deposits of oil which are retrieved using oil rigs sitting on the seabed.

ABOVE:
Offshore wind turbines are becoming more numerous as nations invest in renewable energy sources to supplement existing energy supplies.

Formation and mining of oil

1. Large amounts of plankton debris forms a mud on the seabed rich in organic material.
2. Layers of sediment build up on top of this organic layer, compressing it and heating it until it turns into oil.
3. The oil is lighter than the surrounding rocks and filters upwards as the layers buckle under tectonic forces.
4. The oil accumulates in a reservoir from where a drilling platform mines it.

Sea floor minerals

The floor of the ocean is the home to a wealth of mineral deposits. The growth of the human population and its increase in wealth is putting tremendous pressure on the dwindling supplies of land-based metals. However, the ocean, covering 70 per cent of the planet, is a future source of hundreds, perhaps thousands of years supply of many key metals.

There are approximately ten kinds of currently economic (and hence mined) or potentially economic (hence to be mined in the future) marine minerals. Deposits currently mined include placer deposits, those found in beach or stream environments. Offshore placers are currently a rich source of gold, tin, and offshore diamonds. Offshore sand and gravel supplies one quarter of the building material used in Japan and the U.K.. In the U.S., offshore sand is key for beach replenishment. Various corals, salts and clays are also taken from marine sites for specialty purposes.

Currently large deposits of manganese nodules, manganese crusts, polymetallic sulfides, phosphorites, and methane hydrates are mapped on the seafloor but not yet mined. Manganese nodules contain hundreds of years supply of copper, nickel, cobalt, and manganese. Manganese crusts can provide cobalt, nickel, manganese, and platinum. Polymetallic sulfides can provide gold, silver, copper, lead, and zinc. Phosphorites provide fertilizer. Methane hydrates contain thousands of years supply of gas.

Polymetallic sulfides are the closest of these minerals to being mined. Mining leases have been granted off Papua New Guinea and New Zealand. A mining ship will be operational in 2010 with full scale mining to begin in 2011. This is deep water mining in 3,885–5,180 ft (1,500–2,000 m) and will represent a major industrial shift into the ocean with significant new impact by humans.

ABOVE:
Manganese nodules from the sea floor. The one on the left has been cut in half.

TOP:
Polymetallic sulfides are formed by the expulsion of hot mineral laden water into cold seawater. The minerals are dropped from the volcanic water forming mineral chimneys such as this spectacular hydrothermal chimney at Volcano 19 on the Kermadec Arc, north of New Zealand.

RIGHT:
Cross section view of the base of one of the active chimneys recovered by the ROV Jason II at the Five Towers vent field in the Mariana Arc, Pacific Ocean. The brass-colored mineral lining the central orifice of the chimney is the copper sulfide chalcopyrite.

Telecommunication cables

Have you ever called or emailed someone overseas? That information was most likely transmitted through undersea telecommunication cables. These cables are the fastest way to transmit information between continents and can transmit digitized information equivalent to millions of phone calls simultaneously. Although they were originally used primarily for telephone communication, internet traffic is now twice the volume of that from telephones in some parts of the world.

The first transatlantic telecommunication cable was laid in 1858, and by 2005, these cables linked all continents except Antarctica. The cables are about the diameter of a golf ball, contain optical fibers (usually made from glass) which are surrounded by a steel and waterproof coating, and are installed on the ocean floor.

Installing cables on the ocean floor can impact sensitive ocean habitats and animals, and result in economic conflicts with local fisheries. Cable installation projects are routinely required by governments to consider these issues. Cables can also be damaged from human disturbance from fishing trawlers and anchoring and environmental disturbance from events like undersea avalanches. Overall, telecommunication cables are essential for information transfer and damage to cables can disrupt everyday activities such as banking, internet use, and communication with each other.

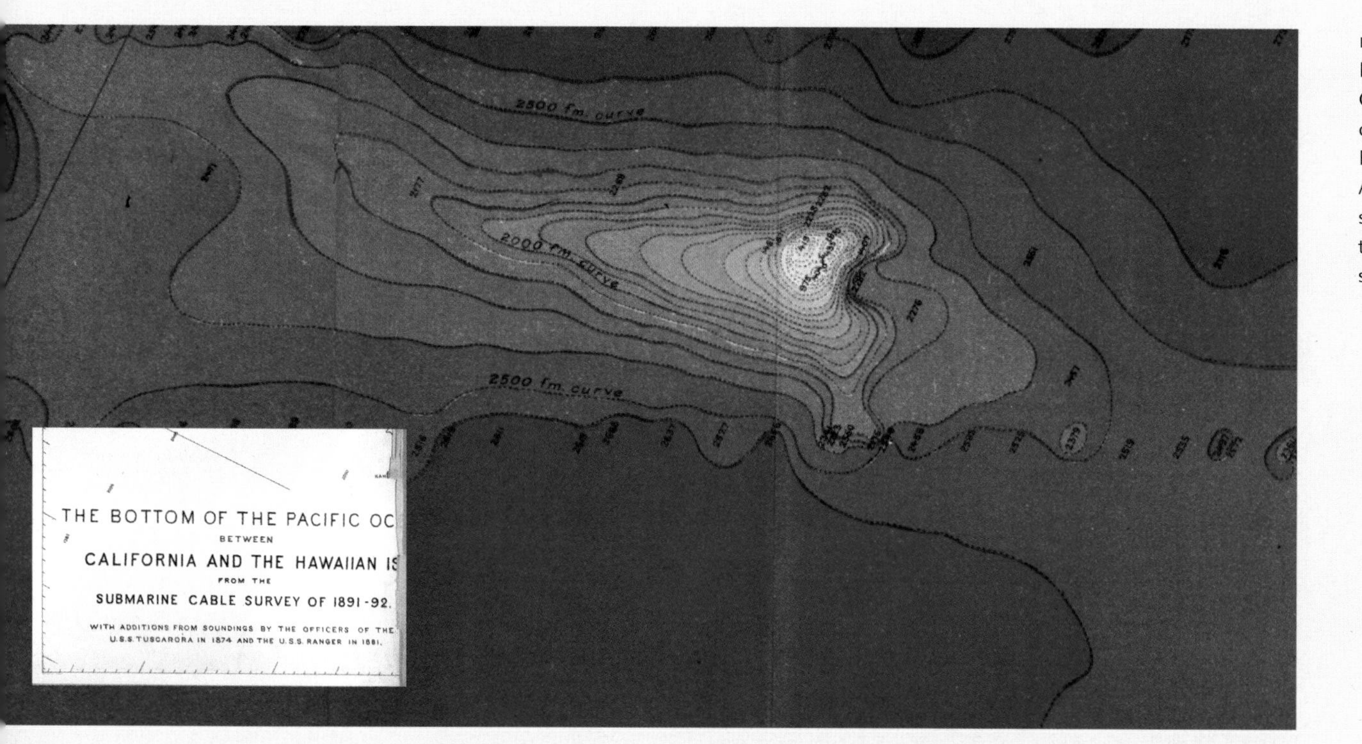

LEFT:
Portion of map in Congressional Report depicting the work of the Fish Commission Steamer Albatross in conducting deep sea soundings for a telegraphic submarine cable survey, 1891–1892.

LEFT:
Early image of landing for submarine cable attached to submarine signaling bell buoy.

Telephone lines and telecommunications traffic worldwide

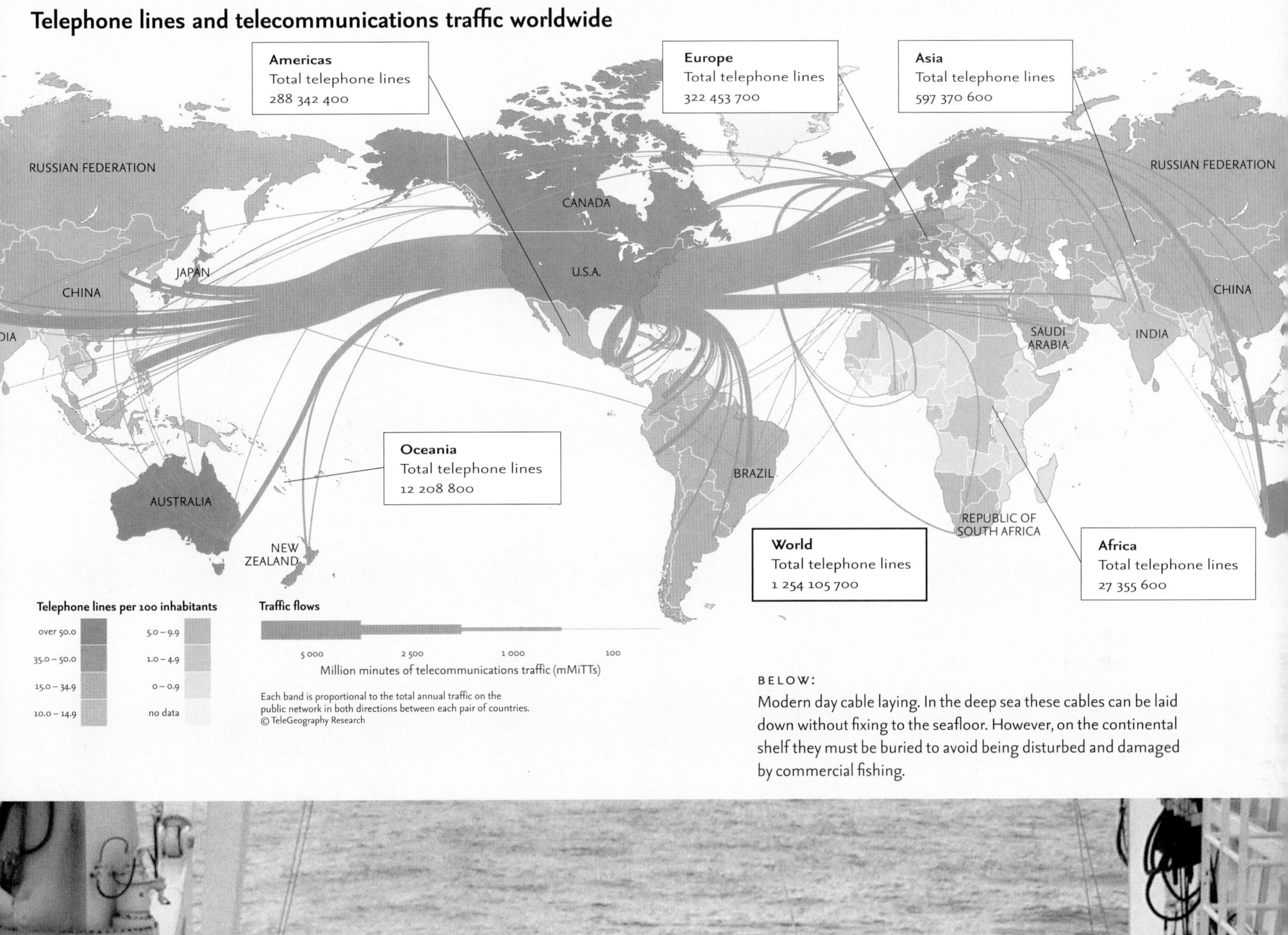

BELOW:
Modern day cable laying. In the deep sea these cables can be laid down without fixing to the seafloor. However, on the continental shelf they must be buried to avoid being disturbed and damaged by commercial fishing.

International law of the sea

ABOVE:
A U.S. Coast Guard boat conducting vessel safety boardings near Chincoteague, Virginia, U.S.A.. Waterways are patrolled in order to keep local boaters safe.

As the international legal framework for activities carried out at sea, the modern-day law of the sea evolved after centuries of navigation, trade, and exploitation. Concerns about depleting oceanic resources, pollution, and security led to a need for balance between a coastal state's right to protect and manage its territory and offshore economic resources while ensuring freedom of navigation and other rights of the high seas, such as fishing, research and the laying of cables and pipelines.

In 1609, Hugo Grotius wrote a treatise called Mare Liberum (Free/Open Sea) which recognized the need for a coastal state to control a narrow belt of waters off its coast as well as the need for freedom of navigation. The "cannon-shot rule" recognized the limit of this belt as the theoretical distance a cannon ball could be fired from shore. As the law of the sea evolved, this narrow belt of sovereign waters would be the territorial sea; however, its exact breadth varied by coastal state.

Following World War I, a movement to codify public international law was underway. In 1945, in response to a national desire to secure the United States' offshore resources, President Truman issued two proclamations to exclusive rights over the continental shelf and to regulate fishing in the high seas area beyond the territorial sea. The proclamations sparked dynamic discussions and new claims which would later establish zones of exclusive control over all economic exploration and exploitation of offshore natural resources through the United Nations Convention on the Law of the Sea (UNCLOS).

By 1982, the custom of asserting a 12 nautical mile (nm) territorial sea, a 200 nm Exclusive Economic Zone (EEZ), and other maritime zones was codified in the UNCLOS. With contributions from more than 150 countries of varying political and socio-economic statuses, UNCLOS contains 320 articles and 9 annexes that preserve freedom of navigation while governing many facets of ocean space including environmental control, scientific research, economic, and commercial activities, and technology.

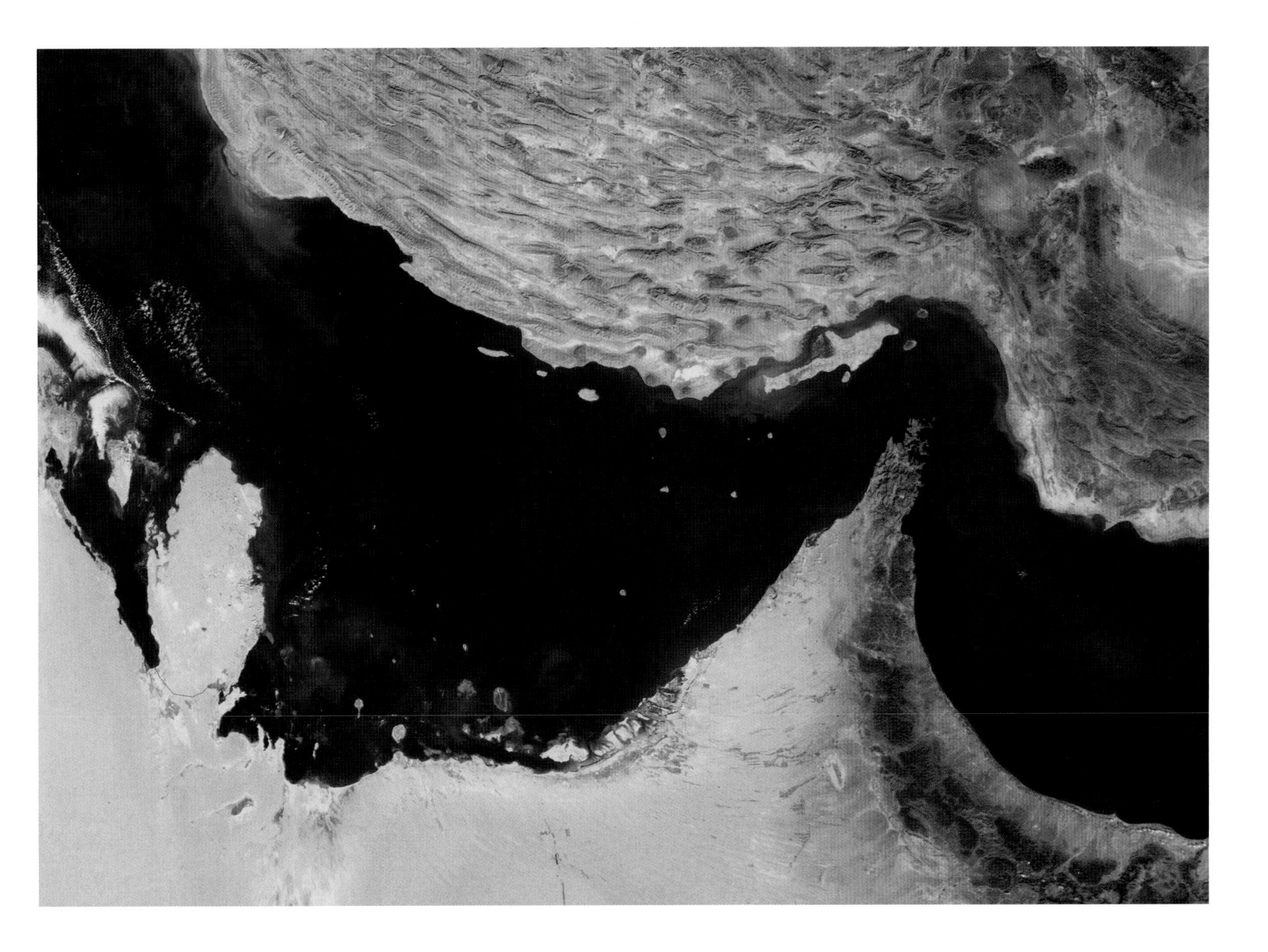

ABOVE:
Right of passage is a major concern of the Law of the Sea. Narrow straits, such as the Strait of Hormuz at the eastern end of the Persian Gulf, provide transit for much of the world's commercial shipping.

ABOVE:
Hugo Grotius, also known as Hugh or Hugeianus De Groot (1583–1645). Dutch jurist and scholar.

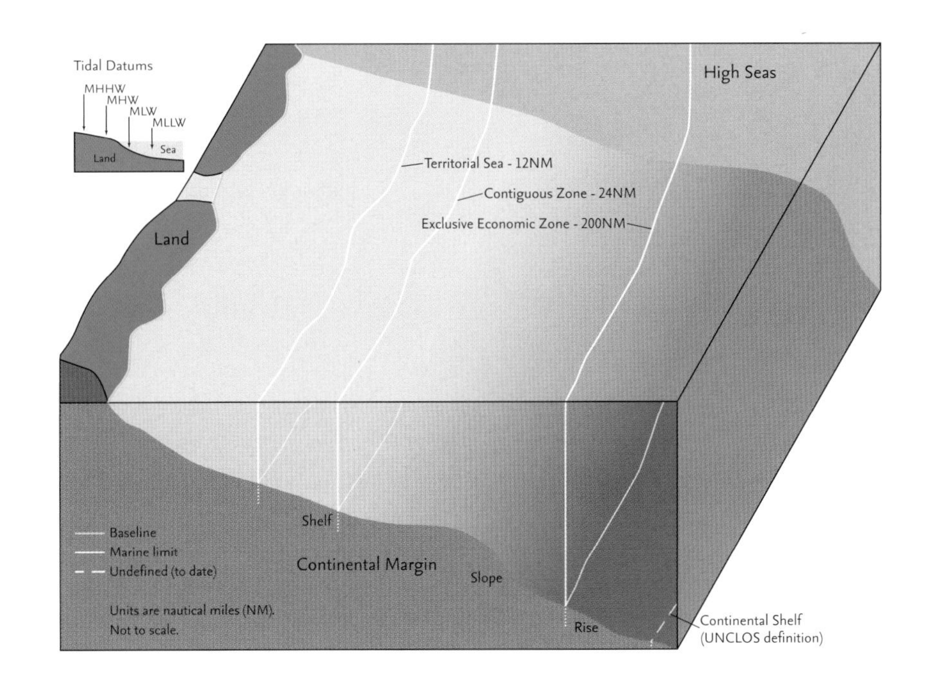

RIGHT:
Maritime zones as measured from a normal baseline. In the U.S., the baseline is the Mean Lower Low Water line as depicted on official nautical charts.

The South China Sea

The South China Sea has more overlapping claims within a single water body than anywhere else in the world. In addition to being a strategically important passageway for transoceanic shipping, the South China Sea has rich fishing grounds and potentially massive oil and gas reserves. In recent decades, disputes over this area have undergone cycles of heated debate, including incidents involving military vessels and garrisons, followed by relative calm as the rapidly developing coastal states seek to secure economic resources for their people.

Sovereignty over features of the Spratly and Paracel Islands, two island groups comprised of small coral islands, rocks, reefs, and submerged banks in the South China Sea, are central to the coastal states' claims to exclusive economic rights of the water column, seabed, and subsoil. To assert sovereignty claims, uniformed personnel and scientists from claimant states occupy about forty-four of the features: China (7) as well as the island of Taiwan (1), Vietnam (25), Malaysia (3), and the Philippines (8). While Brunei has not asserted any claims to islands, it claims an Exclusive Economic Zone (EEZ) and continental shelf which extend out over the southern portion of the Spratly Islands region, incorporating several islands.

To discourage the pattern of armed conflict and casualties in the area, the member states of the Association of Southeast Asian Nations (ASEAN) and China signed a code of conduct in November of 2002 which calls for self-restraint and diplomatic solutions to disagreements rather than force in the South China Sea. Subsequently, in March of 2005, the national oil companies of China, Vietnam, and the Philippines signed an accord to conduct joint oil exploration activities within their overlapping claims in the Spratly Islands. This cooperative work through ASEAN and other joint agreements is a step in the right direction in what is likely to be an enduring dispute.

RIGHT:
Oil rig being towed into position in the South China Sea. The promise of massive oil and gas reserves is a major source of conflict in the region.

BELOW:
Small island and reef in the disputed Spratly Islands.

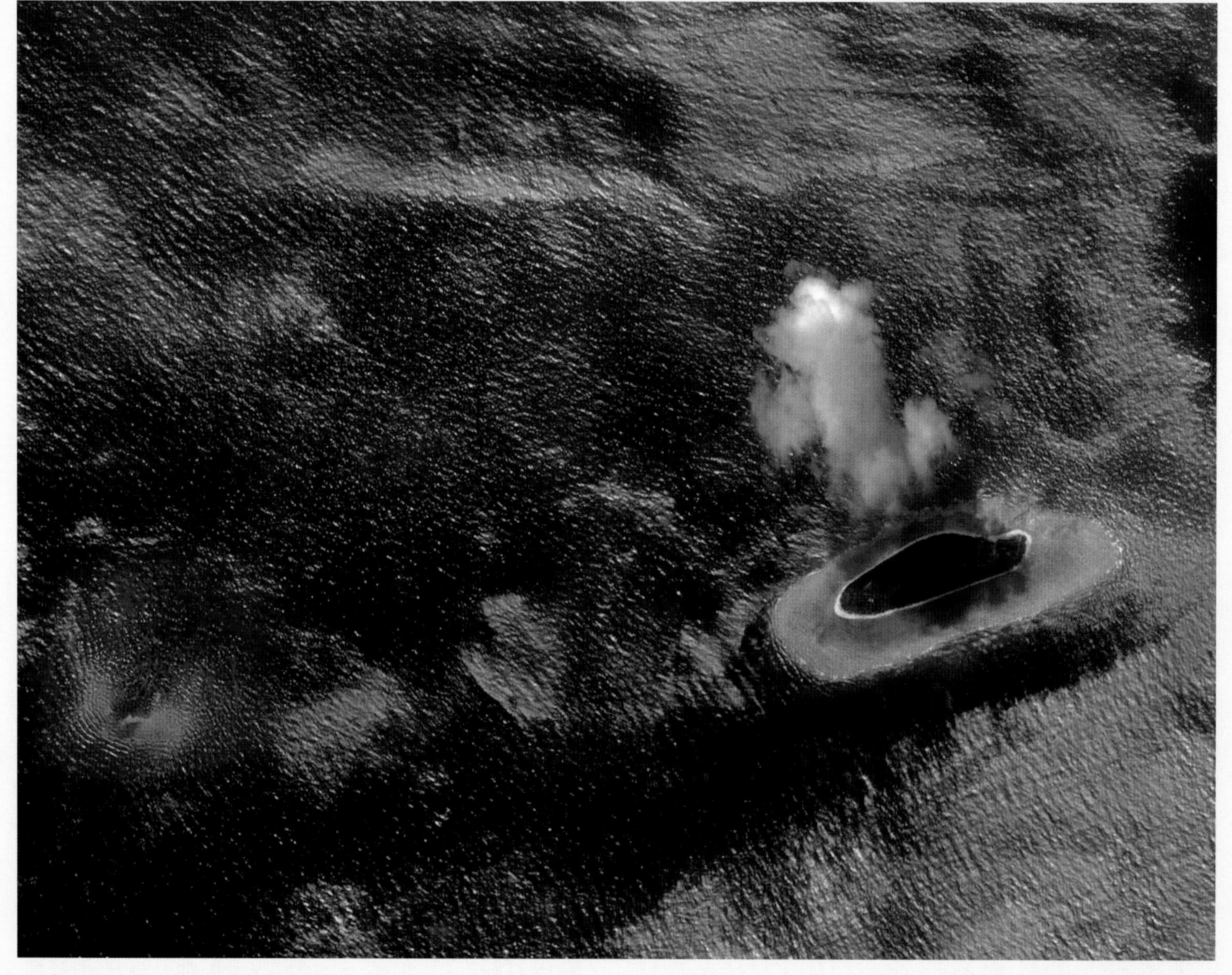

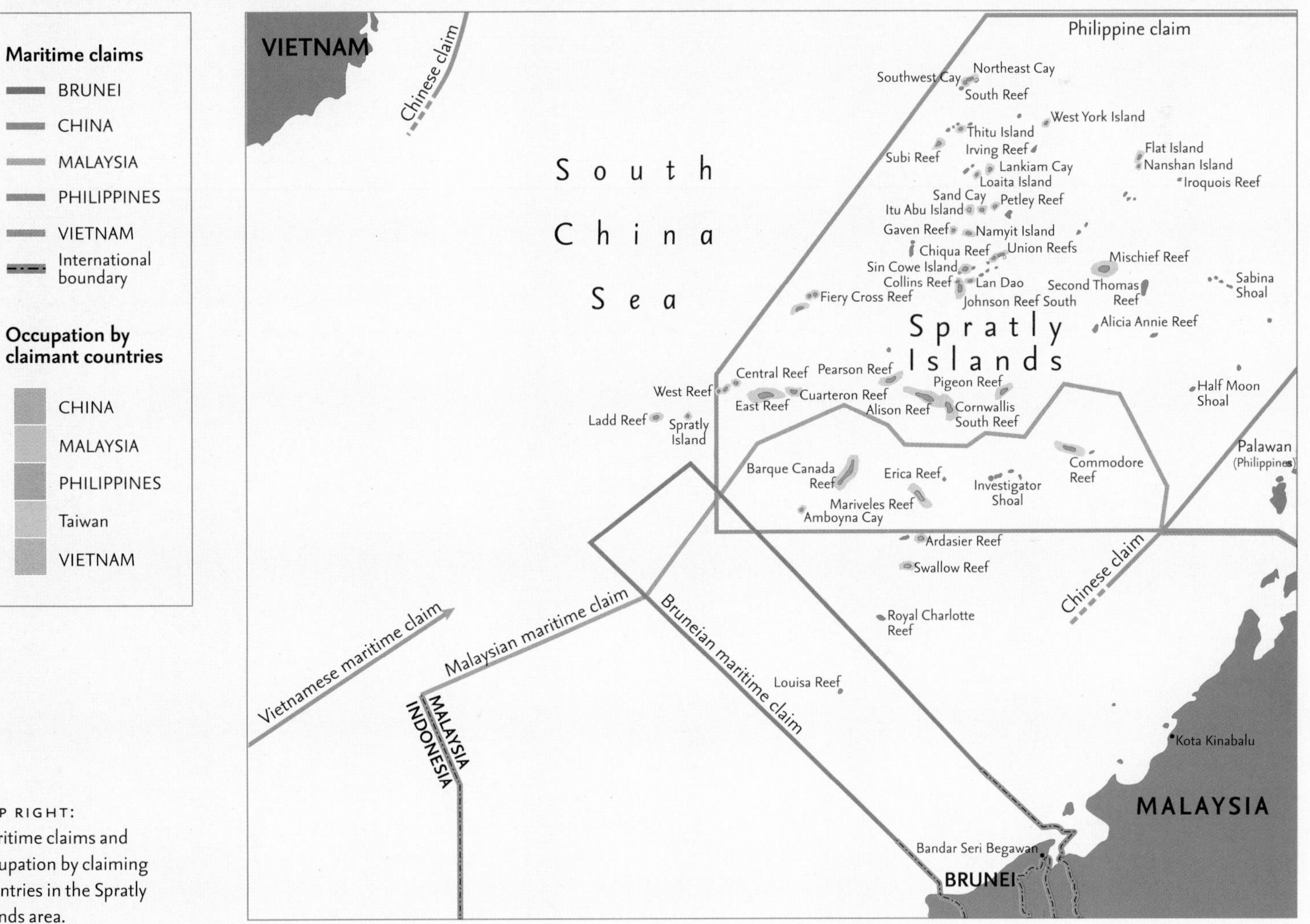

MAP RIGHT:
Maritime claims and occupation by claiming countries in the Spratly Islands area.

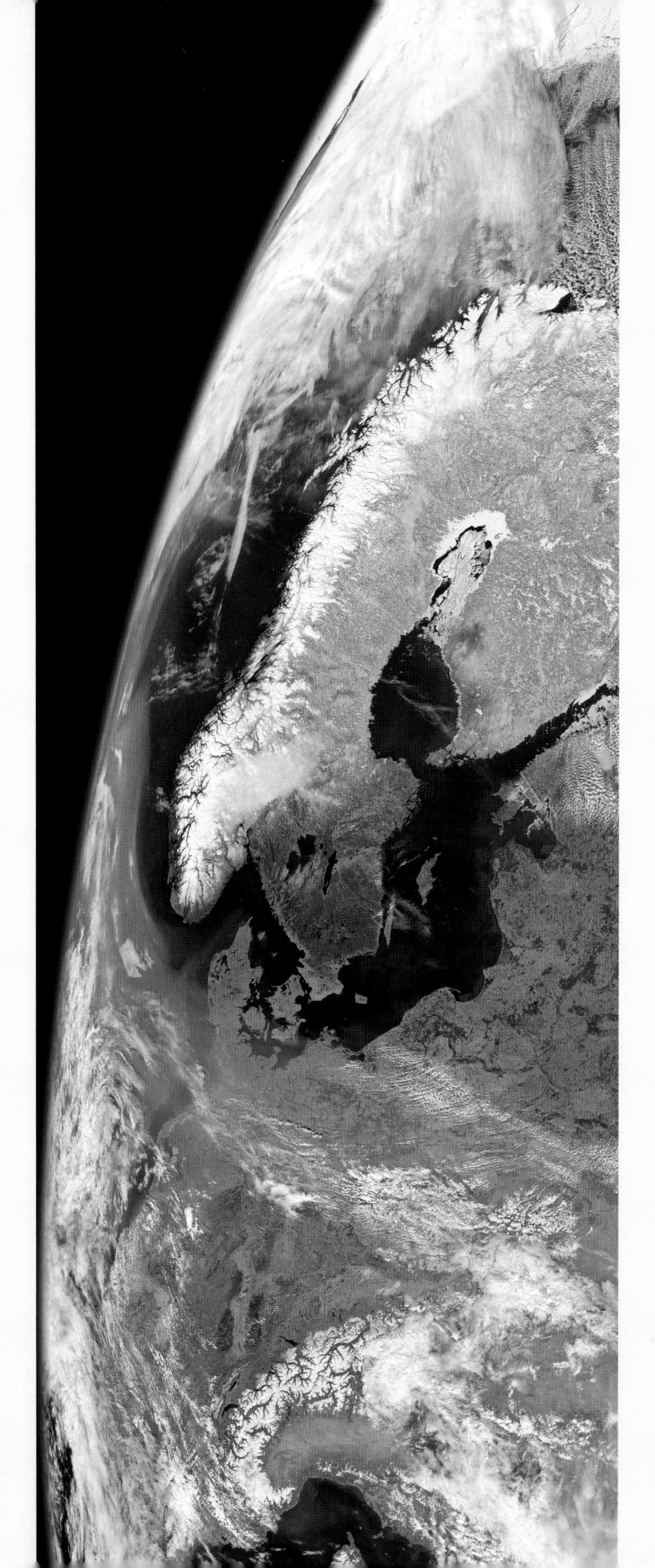

The Baltic Sea

The Baltic Sea, the great inland sea of northern Europe, is one of the youngest seas in the world. Its evolution has undergone changes which have been influenced by human economic activities and by rapid exploitation of natural resources shared by nine countries bordering its waters. The Baltic Sea covers an area of 163,000 sq miles (422,000 sq km), with a relatively shallow average depth of 180 ft (55 m). The greatest depth is about 1,380 ft (421 m) off Gotland Island in the central area. The Baltic Sea is fairly contained and connects with the North Sea only by the Kiel Canal and by the Skagerrak passageway between Norway and Denmark, limiting its flushing capability, and is relatively tideless. It is fed by many rivers and fresh water runoff and its salinity is relatively low. The most northern areas have been partially frozen during very cold winters.

The Baltic has been subjected to man-induced hazards, including oil pollution and thousands of tons of munitions of all kinds which were dumped after World War II. These include chemical weapons which are slowly deteriorating and exposing highly toxic chemicals, such as arsenic and mustard gas, to Baltic fisheries resources and endangering environmental health. More than 1,000 warships and civil vessels sunk in the Baltic Sea during World War II, may have been carrying military cargo and present a potential hazard to the marine ecosystem and its inhabitants. Oil pollution, by accidental spills from tankers and drilling platforms, causes significant damage to the marine environment. Pollution transported by river outflow into the sea is another area of great concern. Energy supply has become a major issue in the region, and there is great concern about a plan to supply gas from Russia to Germany via a pipeline passing through the center of the Baltic along the seafloor. This has the potential of leakage causing environmental impact on a broad scale.

On the positive side, the Baltic Sea provides marine resources, transportation corridors, marine recreation, tourism, and desirable coastal living. The nine countries (Denmark, Estonia, Finland, Germany, Latvia, Lithuania, Poland, Sweden, Russia) bordering the Baltic Sea are cooperating by continuous monitoring of the Baltic ecosystem and sharing research and environmental data in order to detect and assess changes which may impact environmental health. The nine nations are represented in an organization called the Helsinki Commission (HELCOM), which was formed to address and ameliorate marine environmental problems and issues which are common to all. Environmental protection policies are moderate and based on sustainable development approaches. There is a growing awareness of ecological issues. Cleaning the Baltic Sea, preserving biodiversity and monitoring long-

range transboundary pollution are of great importance. Economic benefits are dependent on a clean environment.

The Baltic nations are rich in the seafaring tradition of plying the coastal waters for fisheries and inter-nation shipping and trade. Estimates have as many as 2,000 ships and tankers plying the waters at any time, raising the potential of hazardous spills. The Baltic coastline is approximately 5,000 miles (8,000 km) long, and the principal seaports circling around from west to east are: Copenhagen, Stockholm, Helsinki, St Petersburg, Tallinn, Riga, Klaipeda, and Gdansk. The Baltic nations boast many protected coastal areas which also serve as parks and recreational areas. United Nations Educational, Scientific and Cultural Organization (UNESCO) has declared several coastal areas as World Heritage Sites, including the historic centers of the port cities of Tallinn and Riga and the Curonian Spit shared by Lithuania and Kaliningrad, Russia. The spit is a 60 miles (97 km) long sliver of land separating the Curonian Lagoon, off Lithuania's coast, from the Baltic Sea. It is famous for its natural environment and landscape of prominent sand dunes 130 ft (40 m) high in some areas, along its length. The Baltic region is famous for its amber gemstones, with pieces often washed up on the beaches near Kaliningrad and Lithuania. After 1991, the three Baltic States of Estonia, Latvia and Lithuania gained their freedom from fifty years of Soviet rule. Having become members of the European Union (EU) and North Atlantic Treaty Organization (NATO), they are progressing economically and share in the protection of the marine environment.

ABOVE AND LEFT: Satellite images of the Baltic region in winter (above) and spring (left). In winter, part of the sea area has frozen over but by spring the snow and ice has largely melted in the east, remaining only in Sweden's northern highlands.

Cultural history and traditions

We often regard culture in the historic sense, as people in strange outfits reacting to circumstances or conditions long gone. Tradition is even more shrouded in the distant past, often regarded as irrelevant to life in the modern world.

Yet who we are today, and will be tomorrow, are very much shaped by where we've been. Culture is not static, but rather changing and shifting, shaped by the context in which it occurs. Likewise, traditions evolve as new situations and compelling needs emerge.

Take, for example, the Olympic Coast which has sustained human communities for over 6,000 years and possibly longer. Tribes such as the Makah, Quileute, Hoh and Quinault Nation use the ocean's wealth for food, clothing, transportation, and as inspiration for their vibrant art, music and dance. The seascape both shaped and was shaped by tribal culture.

Those resources were so central to tribal existence and cultural identity that they formed the cornerstone of treaties negotiated in the 1850s between Pacific northwest tribes and Washington Territory's Governor Isaac Stevens. These treaties reserved the tribes' rights to fish, gather shellfish, and hunt at their "usual and accustomed places," creating reservations and support from the federal government in exchange for ceding vast areas of land for white settlement. At the time the treaties were negotiated, marine resources were so plentiful that no one could fathom the stresses which an industrializing world would place on the ocean and its inhabitants.

Go to any tribal community today, and you will see tribal culture in its modern context: a modernized fishing fleet, state-of-the-art science and monitoring programs, salmon habitat restoration projects, and young people creating modern art using ancient symbols. Speak to a tribal leader, and the words of the treaties still ring out as the tribes grapple with tough issues such as dwindling salmon runs. Stand on a wild Olympic Coast beach, witness the welcoming songs and words of respect and place used to greet visiting tribes on a long canoe journey. And feel the past linked to the present and the hopes of the future.

TOP LEFT:
Olympic National Park, Washington, U.S.A..

ABOVE LEFT:
Tribal canoe approaching Neah Bay, home of the Makah Tribe.

ABOVE RIGHT:
Salmon cooked the traditional way fuels canoe pullers on a tribal journey through the Olympic Coast National Marine Sanctuary, Washington, U.S.A..

FAR LEFT:
Tribal canoes approaching the Olympic Peninsula during a long canoe journey.

Conservation

Conservation
Introduction

The sea has long been important to humanity as a source both of food and spiritual value. The concept of conserving its resources has also long been present. Many island communities in the south Pacific still practice, or are resuming, traditional community-based marine conservation, managing areas from the mountains to the sea. Even in colonial times in the United States, area-based conservation was practiced, such as the closure of shellfish beds. However, the conservation of large marine areas and their ecosystems is a relatively new concept, especially when compared to the large-scale terrestrial-based conservation actions which began in the early twentieth century. Today, approaches to marine conservation continue to evolve, becoming more ecosystem-based in approach, and focusing not only on natural resources, but also on the conservation of cultural heritage such as shipwrecks and archaeological sites.

Despite observations to the contrary, such as the disappearance of codfish stocks in the north Atlantic or the diminished number of "big fish" around Caribbean islands, some people well into the twentieth century still spoke of the sea as bottomless and its resources as endless. However, just as the disappearance of large mammal and bird species on land led to their protection through hunting regulations or designation of areas as national parks or refuges with limited or no hunting, the late twentieth century saw increased use of laws and regulations such as gear restrictions to protect or restore fish stocks, seabirds and marine mammals, and the establishment of closed and managed areas, generically referred to as "marine protected areas" (MPAs), to protect or restore species, habitats, and special places from an actual or potential threat.

Discussions today focus on which resource management techniques or combination of techniques can best achieve specific resource conservation goals. Few dispute the need to do so. People also realize that marine resources cannot be protected solely at sea; land-based sources of pollution such as the run-off of stormwater must also be addressed. The ultimate questions for conservationists are: What are we trying to accomplish? How do we get there? And what are the consequences? Considerations include: What is the ecological baseline we wish to re-establish? Is this a feasible and reasonable goal? What are the best ways to address these problems? Are the solutions local, regional, or global? How will this affect the lives of people? Can we monitor the effects resulting from our actions to determine whether or not management measures are working?

There is also growing recognition that many marine resource management problems require multi-national solutions. Early examples include the establishment of international bodies such as the International Whaling Commission to restore whale populations and the International Commission for the Conservation of Atlantic Tunas to manage these highly migratory fish. Marine conservation today often requires not only action by countries individually, but also multi-nationally, especially in "high seas" beyond the jurisdiction of individual nations.

Recently discovered areas of significant ecological value such as deep-water corals and sea mounts are threatened by illegal high-seas fishing practices which are devastating these little understood, and often slow-growing species assemblages and habitats. Large areas of the U.S. Exclusive Economic Zone in the Pacific recently were declared "no trawl" areas to protect such resources. The effect of the loss of these ecosystems on the broader marine environment is yet to be truly known. Most of these places are marginally explored or unexplored and often may be damaged before research has begun or management measures put in place.

PREVIOUS PAGE:
Butterflyfish over corals in the Andaman Sea, Thailand. Those with the yellow tails are panda butterflyfish, while the red-tailed fish are redtail butterflyfish.

RIGHT:
Yellow goatfish shoal above Dolphin Reef in the Red Sea, Egypt. This fish is found throughout the Indo-Pacific region. It generally remains in a shoal during the day, but at night each fish moves off to hunt alone.

RIGHT:
Paul Kroegel (1864-1948). In 1902 Kroegel was hired as warden of Pelican Island, Florida, U.S.A. to protect the threatened brown pelicans which roosted there. With the declaration of the island as America's first national wildlife refuge on March 14, 1903, Kroegel became the first federal game warden.

FAR RIGHT:
Northern or Steller sea lion caught in fishnet.

Marine protected areas

Chances are you've visited a marine protected area (MPA) in the U.S. and don't know it. If you've gone fishing in California, diving in the Florida Keys, camping in Acadia National Park, swimming in Cape Cod, snorkeling in the Virgin Islands, birding in Weeks Bay, Alabama, hiking along Washington's Olympic Coast, or diving on the shipwrecks of Thunder Bay, Michigan, you've probably been one of thousands of visitors to an MPA.

What is a marine protected area? Some people interpret MPAs to mean areas closed to all human activities. Others understand them as special areas set aside for recreation and commercial use, much like national parks. In reality, marine protected area is a term which encompasses a variety of conservation and management methods in the United States. The official U.S. definition of an MPA is: "any area of the marine environment that has been reserved by federal, state, tribal, territorial, or local laws or regulations to provide lasting protection for part or all of the natural and cultural resources therein." In practice, MPAs are defined areas where natural and/or cultural resources are given greater protection than in the surrounding waters. MPAs in the U.S. span a range of habitats including the open ocean, coastal areas, inter-tidal zones, estuaries, and the Great Lakes. They also vary widely in purpose, legal authorities, agencies, management approaches, level of protection, and restrictions on human uses, and are established to protect natural and cultural resources and sustain fisheries.

Currently, there are more than 1,500 MPAs in the U.S., managed by hundreds of federal, state, territorial, and tribal laws. Nearly 100 per cent of the area in U.S. MPAs are "multiple-use" sites in which a variety of human activities are allowed. In contrast, only a fraction of the area in MPAs is "no-take," which prohibits all extractive uses.

LEFT:
Great blue heron on the Atlantic shore of Florida's Guana Tolomato Matanzas National Estuarine Research Reserve, U.S.A.

FAR LEFT:
Looking north to the Sand Beach area from the Otter Cliffs, Mt Desert Island, Acadia National Park, Maine, U.S.A.

LEFT:
Dive tourism in Guam, Mariana Islands. Tourism can place additional stress on reefs. Standing on or handling live corals can damage or even kill the coral polyps.

LEFT:
Diver and the propeller of wooden bulk freighter Monohansett (sunk in 1907), located in the Thunder Bay National Marine Sanctuary, Alpena, Michigan, U.S.A.

Worldwide marine protected areas

Earth's coasts and oceans are stressed. Over half the world's population lives in coastal areas, and that number is expected to increase to 60–75 per cent over the next several decades. Coastal tourism is the fastest-growing segment of global tourism. Half the world's wetlands have been lost. More than half the world's coral reefs are threatened by human activities, and most of the world's major fisheries are overfished. Global marine conservation efforts, including the use of MPAs, are just beginning to catch up to the problem.

While the oceans comprise over 70 per cent of the earth's surface, less than 1 per cent of the marine environment is within protected areas, compared with nearly 9 per cent of the land surface area. Of approximately 40,000 protected areas in the world, only about 3 to 6 per cent are MPAs. Management of these areas is mixed in nature and many MPAs are still "paper parks." The area of the world's oceanic waters which prohibit all resource extraction, including fishing, constitutes less than 1 per cent of the marine environment.

The World Commission on Protected Areas (WCPA) of the World Conservation Union (IUCN) defines a marine protected area as "any area of inter-tidal or sub-tidal terrain, together with its overlying water and associated flora, fauna, historical, and cultural features, which has been reserved by law or other effective means to protect part or all of the enclosed environment." The term MPA is commonly used, but areas identified as MPAs have varying degrees of protection.

The varying definitions and multitude of names (e.g. marine park, marine reserve, marine sanctuary, etc.) add to the debate by causing misunderstanding and uncertainty. MPAs also range from small, highly-protected "no-take" reserves which sustain species and maintain natural resources to very large, multiple-use areas in which the use and removal of resources is permitted, but controlled to ensure that conservation goals are achieved. There is also growing interest in establishing MPAs on the high seas.

ABOVE:
Coral reef in the Glorious Islands, Indian Ocean. This reef supports an abundance of fish including species of sweetlips, fairy basslet and fusilier fish.

LEFT:
Sweeper fish shoal in the Red Sea. These fish are found in large schools near to coral reefs, feeding on zooplankton, tiny microscopic invertebrates.

ABOVE:
A shocking disregard for preserving reefs for the future. A mindless diver carved his name in this coral.

LEFT:
Discarded tire amongst remnants of a dead reef.

Coral reef conservation

Coral reefs around the world are at serious risk from human impacts. Coral bleaching (the process by which a coral colony under environmental stress appears whitened) and ocean acidification associated with global climate change are growing threats. Tourism and recreation, fishing, and coastal development must be carried out responsibly to minimize impacts to coral reef ecosystems. Coral diseases are a serious problem in some areas, often related to poor water quality, coral bleaching, and the cumulative impact of other threats.

These threats present a conservation challenge to protect the thousands of marine species which depend on reefs for habitat, food, and nursery grounds. Hundreds of millions of people around the world directly depend on reefs for food, storm protection, tourism, recreation, medical and chemical discoveries, and to maintain traditional ways of life. Current estimates of the global value of coral reef ecosystem services are as high as U.S. $375 billion annually.

As coral reefs continue to decline globally, reef managers must strengthen efforts to address local threats like land-based pollution and disease. To complement these local-scale initiatives, larger regional and international efforts are needed to manage entire ecosystems and to link marine protected areas across regions and habitat types.

This ecosystem-level approach is exemplified by the recent declaration of 33 per cent of the heavily-used Australian Great Barrier Reef and associated ecosystems as "no-take zones," areas fully protected from the harvest of reef species. The extent and location of the zones were based on extensive monitoring, detailed scientific research on the function of coral reef ecosystems, and thousands of hours of input by the public and coral reef users. After similar levels of public input and scientific assessment of the value of the remote northwestern Hawaiian Islands, in 2006 the U.S. designated this area as a Marine National Monument, the largest marine protected area in the world.

BELOW:
Dead staghorn coral.

BELOW LEFT:
Discarded netting smothers a pristine coral reef.

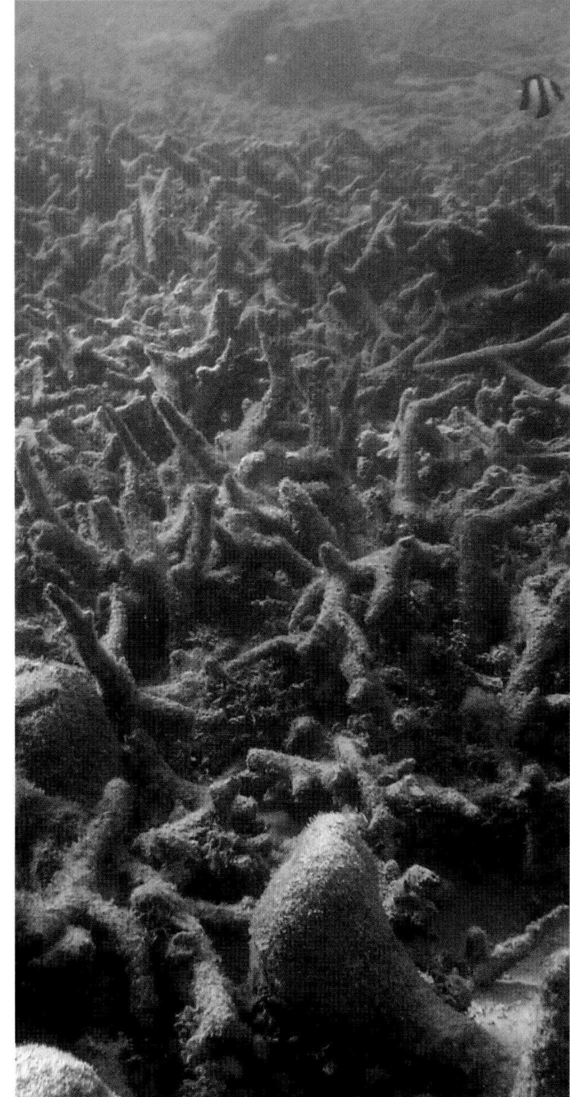

Sustainable fisheries

The ocean has always provided a renewable food source which seemed so abundant that it was considered to be inexhaustible. It wasn't until relatively recently that fishery scientists, utilizing more and better data on fish populations, learned that overfishing of a resource could occur with undesirable consequences to fishing communities and the ecosystem. These scientists found that once a resource was depleted, it was slow to recover without effective management measures and careful monitoring.

Scientific monitoring of marine species in the United States began with the appointment of the U.S. Commission on Fish and Fisheries in 1871. The commission was formed to study the depletion of the cod stock in New England. Since then, the assessment of fish stocks has become an important part of NOAA Fisheries mission. Scientists collect and examine fishery data through a complex modeling process which results in estimates of fish stock population size. These population estimates are used by fishery managers to set harvest limits for individual species to prevent overfishing and allow these stocks to rebuild.

The ability to assess fish stocks continues to advance as new tools and techniques come into play. Deep-sea rovers can take scientists to the deepest fishing areas to examine essential fish habitats. New Fishery Research Vessels contain state-of-the-art research and survey equipment. Vessel monitoring systems utilizing satellites and onboard computers aboard many commercial vessels keep an accurate account of harvest and catch. These data collection improvements are allowing for more accurate assessments and better management of marine species.

In the twenty-first century, we must continue programs which conserve the abundant ocean resources which provide so much to our nation. The U.S. is committed to the long-term sustainability of our marine resources for future generations.

ABOVE:
About 400 tons of jack mackerel are caught by a Chilean purse seiner.

LEFT:
A factory ship in Berkley Sound, Falkland Islands, used to transport the catch from fishing vessels to ports. These ships process the catch before refrigerating or freezing it.

RIGHT:
Frozen tuna fish, numbered for sale in the Tsujiki auction site, Tokyo, Japan.

Protection of marine mammals and turtles

RIGHT:
The green sea turtle is found in warm tropical waters. It is herbivorous, eating algae and sea grasses. Despite the females laying up to 200 eggs at a time, it is an endangered species due to its great number of predators. These include sharks and humans.

Marine mammals and sea turtles face numerous threats which place the survival of individual animals or entire populations at risk. Some threats are: environmental; disease outbreaks; exposure to naturally-occurring marine biotoxins; and extreme weather or oceanographic events (e.g. hurricanes, El Niño). Other threats are linked to human activities: exposure to pollution and contaminants; habitat degradation; entanglement in, or ingestion of, fishing gear and marine debris; and trauma from vessel collisions or intense human-caused sounds in the ocean. In the United States marine mammals and sea turtles are afforded special protection under Federal laws to help their populations thrive, such as the Marine Mammal Protection Act of 1972 or the Endangered Species Act of 1973.

Scientists from government agencies and academic institutions investigate the threats to marine mammals and sea turtles, and develop mitigation measures and conservation plans to reduce impacts at both the individual and population level. Research efforts include collecting and analyzing samples from animals which strand on the beach or are discovered injured or dead at sea; collaborating with the commercial and recreational fishing industries to reduce interactions with the animals; and basic monitoring to assess population status and trends.

For example, NOAA Fisheries Service is involved in cooperative gear research projects designed to reduce sea turtle bycatch in several fisheries throughout the Atlantic Ocean, Gulf of Mexico, and Pacific Ocean. Similar efforts are underway with commercial fisheries and the shipping industry to reduce their impacts on north Atlantic right whales, a critically endangered species. NOAA Fisheries Service also coordinates the marine mammal and sea turtle stranding networks to investigate the causes of strandings and mortality events, which is important for conservation and management of the species, as well as for detection of ocean health issues which can have implications for human health and welfare.

RIGHT:
A humpback whale breaching. This whale is known for these jumps, as well as for its complex vocalizations. A humpback whale can reach around 27 tons in weight and 50 feet (15m) in length. It is found in deep water near coasts in all of the world's oceans.

Red pencil urchin, Papahānaumokuākea

California Sea Lions, Olympic Coast

Spiny lobster, Channel Islands

Elephant seals, Gulf of the Farallones

Gray's Reef sandfish

Installing a Mooring Buoy in Fagatele Bay

The National Marine Sanctuary System

The National Marine Sanctuary System consists of fourteen marine protected areas encompassing more than 150,000 sq miles (388,500 sq km) of ocean and Great Lakes waters from Washington State to the Florida Keys, and from Lake Huron to American Samoa. The system includes thirteen national marine sanctuaries and the Papahānaumokuākea Marine National Monument. Sanctuaries range in size from less than one sq mile to more than 5,300 sq miles (13,720 sq km) while the Papahānaumokuākea monument in the northwestern Hawaiian Islands encompasses almost 140,000 sq miles (362,600 sq km).

The first sanctuary was established in 1975 to protect the USS *Monitor*, a historic Civil War-era ironclad warship. Every site within the sanctuary system has been established because they represent some of America's most important ocean habitats and underwater archaeological sites. Sanctuaries include the Florida Keys, which has the world's third-longest coral barrier reef, and Monterey Bay, which contains a deep-sea canyon deeper and larger than the Grand Canyon. These places are home to an amazing diversity of marine life, including many endangered species.

The National Marine Sanctuary System, part of the National Oceanic and Atmospheric Administration, manages sanctuaries by working cooperatively with the public to protect marine life and habitats while allowing compatible recreation and commercial activities. The sanctuary program works to enhance public awareness of our marine resources and marine heritage through scientific research, monitoring, exploration, educational programs, and outreach. Following evolving international practice for marine protected area (MPA) management, the National Marine Sanctuary Program practices what is called "adaptive management" by monitoring site performance, determining the effectiveness of management measures, reviewing and modifying sanctuary management plans, making changes where necessary to improve conservation and address emerging resource management issues.

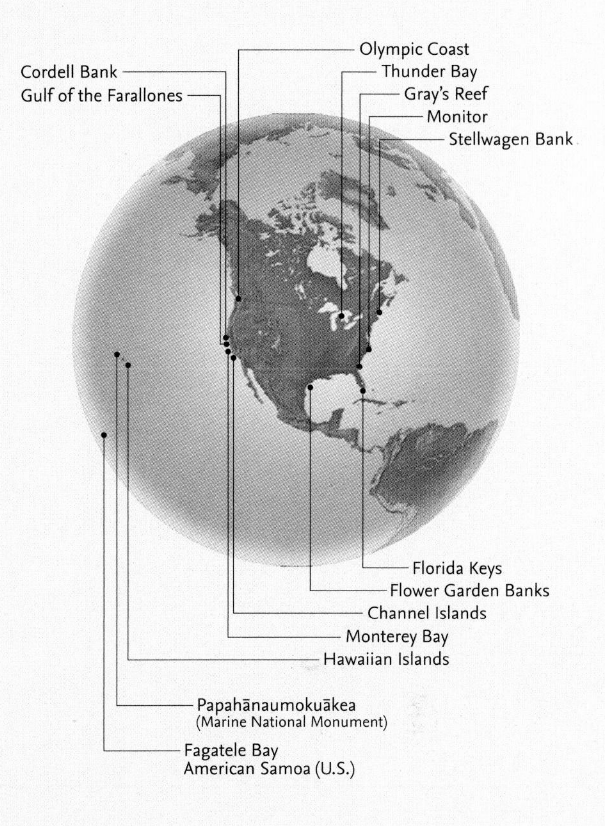

BELOW:
Monterey Bay jellyfish

The Future

The Future
Introduction

In the decades and century to come, mankind will experience – and thus benefit and potentially suffer from – dramatic changes in the nature of our world's oceans, and the impact they have on our lives. Our ages-old reliance on understanding the dynamism of the ocean's physical, biological, chemical, and geological character will not change, but with an increase in that understanding, our opportunities to prepare for, and even exploit, that knowledge will improve dramatically.

In general, that increased understanding will come in two categories: knowledge regarding resources, and skill in forecasting change.

The resources of the oceans (both living and non-living) are extraordinarily diverse. We will, in the future, develop a much-improved understanding of not only what the resources are, but what the limitations are on their availability. In that regard, our consideration of the oceans as a more viable source of food, energy, biomedical products, and materials will grow dramatically, as will our understanding of how to harness these resources in a sustainable or renewable fashion. Just as new energy sources, such as deposits of methane in the ocean bottom, were unexpected and unreported until very late in the twentieth century, we should expect additional discoveries of valuable products and materials from the sea in the years to come. Our continually improving understanding of society's need for conscientious management of limited resources will help in our utilization of these resources for the benefit of mankind, in an environmentally sound framework.

Similarly, in the future we will reap the benefits of dramatically-improved capabilities to predict changes in our oceans, from the physical dynamics (e.g. strength of major current components, such as the Gulf Stream) to the intricate nature of the food web. In the late twentieth century we learned about the chaotic nature of Earth's systems (weather, for example), and we made great strides in understanding the linkages between the atmosphere, the oceans, and the terrestrial environment. As a result we could make well-quantified improvements in our abilities to forecast such important phenomena as hurricanes. How will we extend these wonderful predictive capabilities in the oceans?

We've learned of the general nature of "teleconnections," that is, the fact that features in one part of the ocean (such as the temperature of the surface waters in the equatorial Pacific during an El Niño event) might affect features in other parts of the world (such as drought in parts of Africa). The challenge we will meet in the years to come will be to develop a robust and accurate means to understand these causes and predict the effects well in advance, so that decision makers can adjust accordingly. But these predictive capabilities will not be limited to the purely physical characteristics. Indeed, we will develop

PREVIOUS PAGE:
Computer artwork of a long distance underwater tunnel. The tunnel, which is neutrally buoyant, remains secured in place by vertical tethers attached to floating pontoons . Transport such as maglev trains could travel at high speed without stopping until reaching their destination. A journey from London to New York travelling at 311mi/hr (500 km/hr) would take 11 hours. If built, tunnel engineers must overcome the difficulty of building the tunnel in mid-ocean. It must also withstand the enormous pressure of water at depth and overcome the forces of ocean currents and storms.

the skills to forecast those very complicated processes involving living resources as well: both the detrimental, such as a harmful algal bloom (i.e. red tide) and the beneficial, such as the increased stocks of commercially-viable fish species.

So how will we meet this future world of diverse resources and improved forecasts? Much of this will be possible by dint of exciting improvements in our ability to observe the oceans. Current efforts, for example, to build and sustain ocean observing systems, which can work hand-in-glove with our weather observations, and environmental satellite systems, will provide a revolutionary capability for quantifying the nature of the ocean's physics, chemistry, biology, and geology. Much as the advent of radar as a weather observing tool yielded dramatic improvements in our understanding and exploitation of weather, the same will hold true for our growing capabilities for observing the oceans.

In the future the oceans will be an increasingly important engine for economic and social development. Every generation has benefited from an improved understanding of the oceans. From the earliest explorers who optimized their navigation through knowledge of the ocean currents, to the future society who will better manage energy, food, and water through improved ocean forecasts, mankind has stood to gain dramatically from knowledge about the oceans.

LEFT:
(from the top)
A flotilla of fish follow a transparent drifting jellyfish.
Flame scallop.
Sponge, coral, and searod.
Caribbean spiny lobster.

RIGHT:
(from the top)
Humpback whale.
Research ship in sea ice in the Arctic Ocean.
Diver exploring a World War II era shipwreck near Rota, in the Commonwealth of the Northern Mariana Islands.
Coral.

FAR RIGHT:
(from the top)
Tufted puffin.
A tropical storm in the western Atlantic Ocean.
Icicle.
Pacific atoll from space.

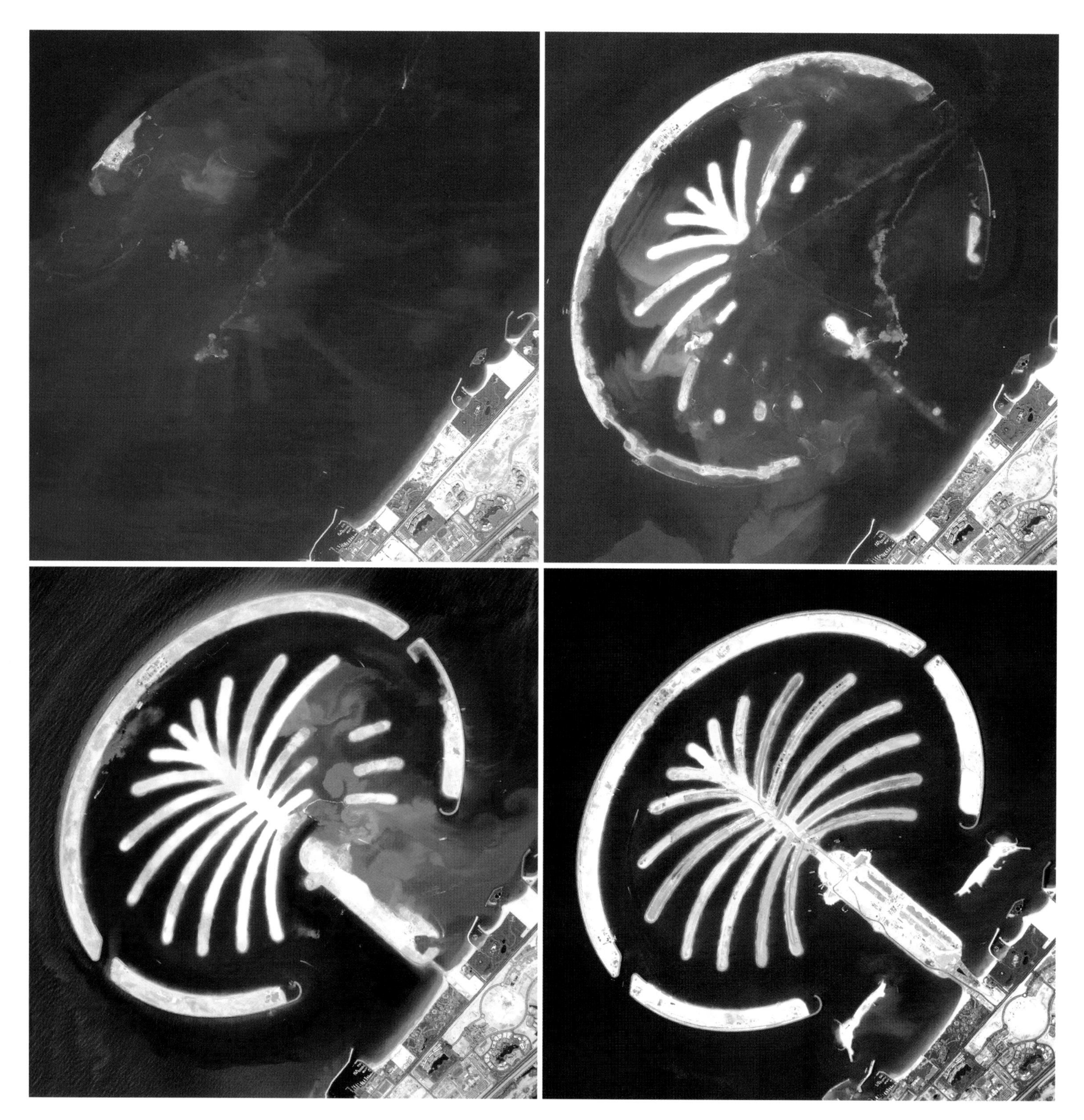

Sand and gravel

Sand and gravel resources are essential to all coastal communities as materials to construct roads, buildings, and also to renourish beaches. Many coastal communities are faced with coastal erosion and land loss, as well as potential impacts from climate change and accelerated sea-level rise, with increased coastal vulnerability to erosion and flooding.

Beach nourishment is increasingly used as a means of mitigating coastal erosion, providing flood protection, and restoring degraded coastal ecosystems. However, large volumes of high quality sand are needed for a beach nourishment project to be successful. Even larger volumes of sand and gravel are used for land reclamation projects in places like Hong Kong and Singapore, both of which continue to extend their coastlines to allow for new developments. Marine sand bodies on the inner to mid-continental shelf are attractive targets for sand dredging, although the geologic character and distribution of sand bodies are often highly variable depending on the complex shelf history and processes of marine transgressions over the past 20,000 years.

ABOVE:
The first "Palm Island" taking shape on the coast of Dubai, United Arab Emirates, in February 2002, November 2002, June 2003, and September 2004. This is the world's largest land reclamation project to create a series of man-made islands on which to build luxury resorts. More of these islands are either already under construction or planned for the future.

Seaweed

Seaweed has been harvested in Europe, Asia, and the Pacific Islands for thousands of years, and is now big international business. Not only are seaweeds harvested for direct consumption of the plant itself, but also for the functional chemicals or "natural products" they produce. Many of these compounds have application for human use. Chemicals derived from seaweeds are used in medicines, food and beauty products, and industry. Most seaweeds are algae, aquatic plants which lack roots, stems or leaves, and are divided into three main types: red, brown, and green. Red and brown algae produce phycocolloids ("phyco" = seaweed, "colloid" = glue) which include agar, alginate, and carrageen. Green algae produce the antioxidant beta-carotene.

The use of these compounds in food products took off in the second half of the twentieth century as the demand for prepared foods increased. Compounds like carrageen improve the quality of the food and help to stabilize it, making the item more appealing to consumers. Currently the import and export of seaweed worldwide has a value in the order of $500 million. As more nations become developed, the need for more prepared foods and pharmaceuticals will increase the demand for seaweed compounds. To meet this demand, selected marine algae are grown, harvested, and processed on a large scale around the globe.

RIGHT:
A patch of dark green seaweed exposed at low tide. The swollen, round structures are air bladders, which keep the seaweed buoyant when submerged. The tough leathery fronds, which are anchored to rocks in the intertidal zone of the shore, can survive repeated bashing by waves during rising tides. Seaweeds belong to a primitive group of plants called algae. Like other algae, they lack leaves, roots, flowers and veins, but like all plants, they can photosynthesize (use the sun's energy). Seaweeds have found a variety of commercial uses from human food to production of the jelly-making substance agar.

Biotechnology

Marine biotechnology is a relatively new field which involves the discovery and application of products and processes derived from marine organisms. Marine biotechnology's promising future reflects the tremendous biodiversity of the world's oceans and seas. As the oceans and seas cover more than 70 per cent of the earth's surface, marine organisms account for a major share of the planet's biological resources. Most major groups of living organisms are primarily or exclusively marine. Science has identified close to 200,000 species of marine algae, animal bacteria, fungi and viruses, with perhaps as many as four times this number of organisms as yet unidentified. Tropical marine environments harbor an especially wide diversity of animals and plants.

The promise of marine biotechnology also reflects many marine organisms' need to adapt themselves to the extremes of temperature, pressure and darkness which are found in the world's seas. The demands of the marine environment have led these organisms to evolve unique structures, metabolic pathways, reproductive systems, and sensory and defense mechanisms. Many of these same properties have important potential applications in the terrestrial world.

BELOW:
Blue mussels, used in the research into new adhesive materials.

For example, many marine organisms are sedentary and must employ sophisticated methods to compete for a place to anchor. Barnacles and mussels, which depend on their ability to attach to solid surfaces for survival, have developed bio-adhesives that stick to all kinds of wet surfaces. Research into the ways which marine organisms adhere to wet surfaces, or prevent other organisms from adhering to them, is yielding useful new technologies. These technologies include both adhesion inhibitors (e.g. anti-fouling coatings for ship hulls) and new types of adhesive such as medical "glues" for joining tissue or promoting cell attachment in tissue engineering applications.

About half of all the drugs currently in use are derived from natural products. These include many of the anti-infective and anti-tumor drugs developed during the past twenty years. Terrestrial plants have been the traditional source but marine organisms are the most rapidly expanding sector. There is nothing new in using marine products for medicinal purposes, with seahorses being used for centuries in traditional treatments for sexual disorders, respiratory and circulatory problems, kidney and liver diseases, amongst other ailments, in China, Japan, and Taiwan.

More recent research has developed a cancer therapy made from algae and a painkiller derived from the toxins in cone snail venom. Anti-viral drugs and anti-cancer agents have been developed from a Caribbean coral reef sponge, and a statin, extracted from an Indian Ocean sea hare, is undergoing clinical trials for the treatment of breast cancer, tumors, and leukemia.

At first organisms tended to be collected in huge quantities, more or less at random, in the vain hope that some useful compound might be extracted later. Genetic engineering has made prospecting for new drugs much more environmentally friendly, it being routine to collect as little as 2 lb (1 kg) of living material. DNA is then extracted from this and cloned into host bacterial cells which produce large quantities of the chemical in the laboratory.

In spite of the sea's vast potential as a source of new drugs and biotechnologies, this domain remains relatively unexplored. Relatively few marine biotechnology products and services have been commercialized to date. Indeed, the vast majority of marine organisms (primarily micro-organisms) have yet to be identified.

RIGHT:
The deadly venom of the cone shell, a type of marine snail, contains proteins (conotoxins) which include the amino acid Gla (gamma-carboxyglutamic acid) in their structures. This is involved in the blood clotting process in humans. It is hoped that the study of the biochemistry of cone shells could lead to the development of new drugs to treat clotting disorders such as haemophilia.

RIGHT:
Seahorses have been used for centuries in traditional medicines.

Future of fish

Fish have a unique role in the ocean's ecosystems. As the targets of commercial fisheries, they provide employment, opportunity, and wealth to coastal communities (including indigenous cultures), and seafood products to all. Recreational fishing and fish viewing is enjoyed by millions and further enhances the well-being of coastal communities. Fish themselves are important components of the ecosystem. Smaller fish are prey to larger organisms, including marine mammals and birds; larger fish help structure the ecosystem's food web through their predation on smaller organisms. These diverse roles for fish pose a fundamental challenge: how can we obtain benefits from harvesting fish while maintaining the benefits of leaving fish in the ocean? What is the optimal amount to catch and who decides? The stewardship mission of the National Oceanic and Atmospheric Administration (NOAA) and similar agencies worldwide focuses on providing the knowledge and leadership to answer these questions.

In the U.S., tools to determine and implement the right level of fishing have never been stronger. In 2006, the reauthorization of the Magnuson-Stevens Fishery Conservation and Management Act now requires implementation of annual catch limits, based on the best scientific information, in all U.S. fisheries to prevent overfishing. It provides for establishment of limited access privilege programs to enhance the safety and economic viability of commercial fisheries and improves measurement of catch by recreational fisheries. Inclusion of ecosystem considerations in fishery management decisions is fostered, international arrangements are enhanced, and cooperative research programs are expanded to provide needed data. Simultaneously, NOAA is using the Ocean Action Plan to increase its ability to conduct integrated ecosystem assessments which will complement fishery management plans. Globally, there is increasing attention to eliminating illegal and unreported catch and protection of sensitive fish habitats from harmful fishing gears.

Tools to manage fisheries to a sustainable level of catch must be supported by scientific programs to calculate this level and track its fluctuations over time. Each year, fishery agencies monitor the catch in commercial and recreational fisheries, deploy observers to monitor bycatch, conduct surveys and biological studies using dedicated fishery research vessels and other chartered vessels to monitor trends in the abundance and distribution of fish populations, and then meld these data into increasingly comprehensive models to update assessment reports.

ABOVE:
Farmed sea bass in a cage off the coast of Corsica. Sea bass is a prized food fish of the northern Atlantic Ocean and Mediterranean Sea. Aquaculture of such fish allows producers to control supply to meet demand, and farmed fish are often fatter than wild fish.

The quality of these reports continues to increase as modern, quiet vessels enter the research fleets, fisheries are more completely monitored, new technologies let us measure fish abundance in previously unreachable habitats, and the models begin to integrate ecosystem and environmental factors. These assessment reports provide the information needed to detect overfishing and to guide fisheries to sustainable levels. From a broader perspective, monitoring of fish stocks will provide leading information on the whole ecosystem's response to climate change, fishing, and the myriad other factors which influence the ocean's ecosystems. With this information and the tools to adjust fishery management, the future of fish and their ecosystems can be good.

In addition to wild capture fisheries, aquaculture – also known as fish and shellfish farming – is playing an increasingly important role in seafood production and species replenishment worldwide. It's no secret that fish and shellfish farming are integral to global seafood production. According to the United Nation's Food and Agriculture Organization, nearly half of all fish consumed today are farmed, not caught. In the U.S. currently, just over $1 billion of seafood is farmed under stringent regulations that protect water quality and ensure aquatic animal health. The majority of U.S. aquaculture is freshwater farming of finfish, while the dominant marine aquaculture industry is shellfish, especially oysters and hard clams.

With the demand for seafood on the rise here and abroad, and a seafood trade deficit of almost $9 billion, the U.S. is beginning to take a much more aggressive look at expanding domestic aquaculture as a necessary and important complement to wild caught fisheries. Coastal communities, including fishermen, already play a major role in coastal aquaculture operations in the U.S. In fact, aquaculture plays a significant role in many commercial marine fisheries, including Alaska's, where hatchery-produced salmon make up 20 to 40 per cent of the catch annually. U.S. fishermen are also among those who are successfully pioneering finfish farming in the open ocean in Hawaii, Puerto Rico, and New Hampshire and mussel culture in New England.

Because wild harvests can no longer keep up with growing demand, increases in the seafood supply will come from aquaculture. In today's world even the best managed wild fisheries can't meet the growing demand for seafood.

New ocean resources

ABOVE:
Offshore wind farm at sunrise. These wind turbines have been built off the coast to take advantage of coastal winds to generate electricity.

BELOW:
The top of a black smoker chimney on the southeastern edge of the Iguanas vent field in the Galapagos Islands.

While there has been a substantial increase in renewable and non-renewable coastal and ocean energy projects, there has been a lack of coordinated management strategies to address regional energy needs and siting and impact issues. In the U.S., energy security, the high price of oil, and climate change are fueling emerging renewable ocean energy technologies. Wind farms are proposed offshore at Massachusetts, Long Island and Texas. Tidal power projects (underwater turbines in areas of high currents) are being proposed for coastal rivers in New Hampshire, New York, Oregon, San Francisco Bay and other high tidal/current areas. A proposal to place a mega-turbine on its side within the Gulf Stream off the coast of Florida has been discussed for several years now. Ocean wave energy projects are anticipated in the Pacific northwest. Ocean Thermal Energy Conversion (OTEC), first proposed in the 1980s, is re-emerging as an alternative to derive energy from the thermal gradient differences in deep ocean areas upwelling into shallower areas.

OTEC may one day provide a means to mine ocean water for trace elements. One element in sea water which has historically been sought after is gold. There are vast amounts of gold dissolved in sea water, but the current cost of extracting it is prohibitive. Most economic analyses have suggested that mining the ocean for any valuable dissolved substances would be unprofitable because so much energy is required to pump the large volume of water needed and because of the expense involved in separating the minerals from seawater. But with OTEC plants already pumping the water, the only remaining economic challenge is to reduce the cost of the extraction process. The Japanese recently began investigating the concept of combining the extraction of uranium dissolved in seawater with wave-energy technology. They found that developments in other technologies (especially materials sciences) were improving the viability of mineral extraction processes which employ ocean energy.

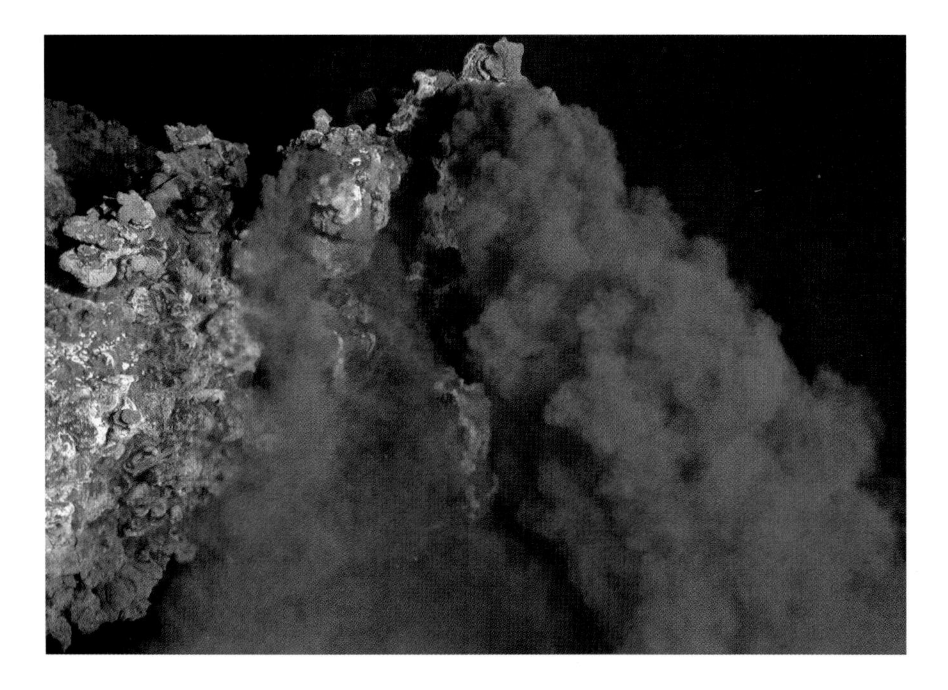

Black smokers are a type of hydrothermal vent found on the ocean floor. Also known as sea vents, these are formed in fields hundreds of yards wide when very hot water comes through the ocean floor. This superheated water is rich in dissolved minerals from the crust, most notably sulfides, which crystallize to create a chimney-like structure around each vent. When the superheated water in the vent comes in contact with the cold ocean water, many minerals are precipitated, creating the distinctive black color. The metal sulfides which are deposited can become massive ore deposits over time.

Hydrothermal vents were first discovered in 1977 around the Galapagos Islands using a small submersible vehicle. In 1979, very high temperature hot springs known as black smokers were discovered on the East Pacific Rise near the entrance to the Gulf of California. Today, black smokers are known to exist in the Atlantic and Pacific Oceans, at an average depth of 6,900 ft (2,100 m). The temperature of the water at the vent can reach 752°F (400°C), but does not usually boil at the seafloor due to the high pressure it is under at that depth. Each year 3×10^{14} lbs (1.4×10^{14} kg) of water is passed through black smokers.

Methane

RIGHT:
Methane bubbles trapped in ice.

The sea's resources are quite diverse, and as our skills for exploration and knowledge increase we continue to expand the potential value of these resources. In the late twentieth century, with the advent of sophisticated acoustic technologies, scientists discovered deposits of "frozen gases" within the depths of the sediments of the world's oceans. These deposits, known as methane hydrates, are not actually frozen gases, but are a unique class of chemical substance in which molecules of one material (in this case, water) form an open solid lattice which encloses, without chemical bonding, appropriately-sized molecules of another material (in this case, methane). The locations, sizes and full chemical construct of these deposits are just now being defined, but as research is conducted we are learning that these features appear all over the globe, and may represent vast energy sources.

However, with the potential benefit of energy supply come the concerns associated with understanding the environmental role played by methane deposits. Since methane is a recognized greenhouse gas, what are the implications of tapping, or inadvertently releasing these hydrate deposits? The size, location, and structure of methane hydrate deposits must also be considered in light of their role in stabilizing the seabed, and in contributing to the ecosystem (many of these methane hydrate deposits exist in conjunction with methane-consuming organisms which play an important role in the foodweb). With these concerns in mind, it is exciting to think about the potential role which methane hydrates might play in our future.

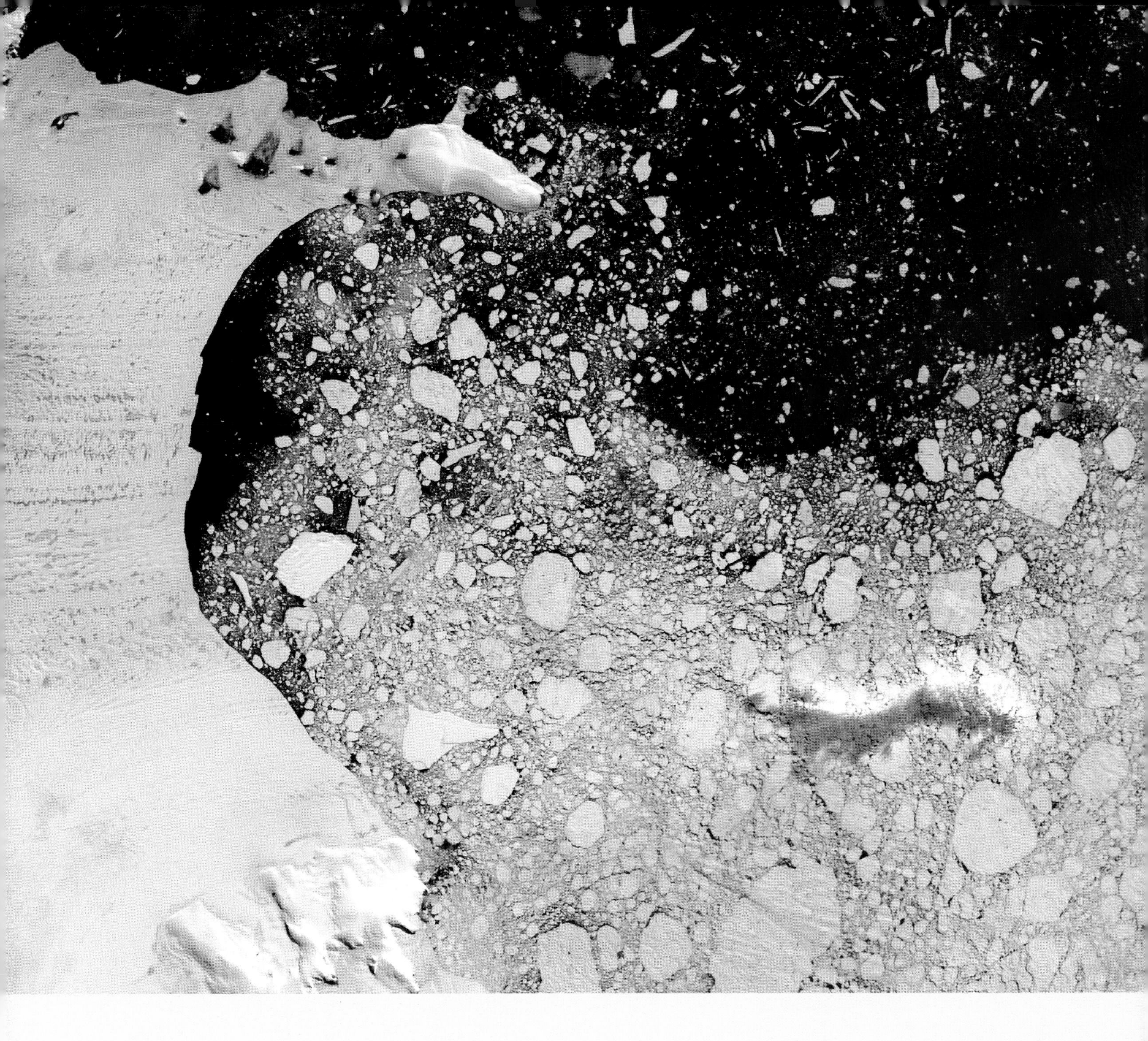

Global warming

"Warming of the climate system is unequivocal, as is now evident from observations of increases in global average air and ocean temperatures, widespread melting of snow and ice, and rising global mean sea level" (Intergovernmental Panel on Climate Change, (IPCC) 2007). In this report, the international community of scientific experts confirmed that since 1961, the average temperature of the global ocean has increased to depths of at least 9,840 ft (3,000 m) and that the ocean has been absorbing more than 80 per cent of the heat added to the climate system. Furthermore, all indications are that warming would continue for centuries even if greenhouse gas concentrations were to be stabilized.

What does such a change in the ocean's temperature mean? The ocean physics and ecosystems are very sensitive to seemingly small changes in temperature. Even average temperature changes of tenths of a degree Celsius will precipitate dramatic change in how heat is distributed through the ocean and how the food web behaves. Ocean currents will be affected both in magnitude and position. Since they drive much of our weather, those patterns will also undoubtedly change. Similarly, the geographical construct of biology in the oceans will change. Migrating species may move to different ocean zones.

Those organisms (e.g. corals) which are fixed in their environment may no longer be able to tolerate the thermal changes, and will either adapt or disappear. And associated with changes in temperature we may see chemical changes, such as changes in the acidity of the oceans as well.

For mankind, we must develop the ability to observe and forecast change, in both space and time, in a manner that will allow well-informed decisions regarding our policies and practices in response to global warming.

LEFT:
The Larsen ice shelf, at the northern end of the Antarctic Peninsula, experienced a dramatic collapse between January 31 and March 7, 2002. First, melt ponds appeared on the ice shelf during these summer months then a minor collapse of about 310 sq miles (800 sq km) occurred. Finally, a 1,000 sq mile (2,600 sq km) collapse took place, leaving thousands of sliver icebergs and berg fragments where the shelf formerly lay.

BELOW:
The changing Arctic. In the Arctic, temperatures have risen at twice the global average rate over the last few decades. All indicators suggest that warming will continue and it is projected that the southern limits of permafrost, the treeline, and summer sea ice for example, will gradually move north.

Combined global land and sea surface temperatures 1856–2004

Relative to 1961–1990 average. The orange line is a smoothing of the annual values.

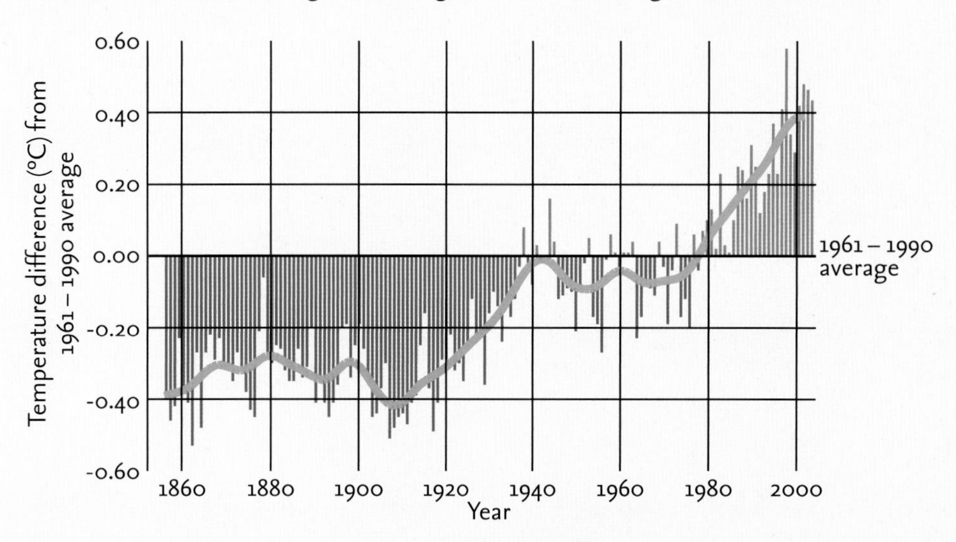

1885

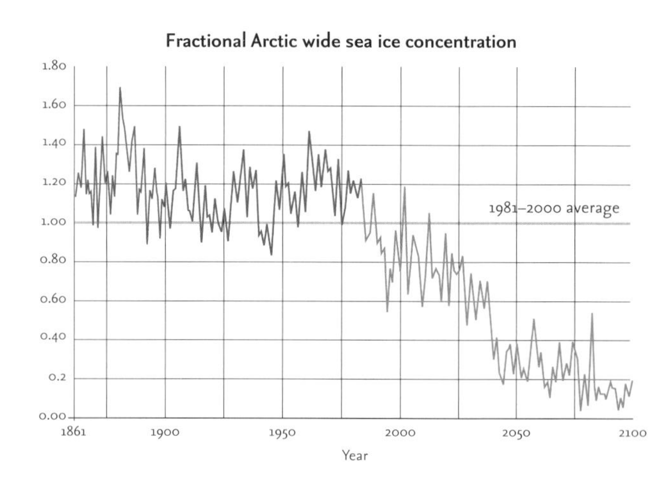

ABOVE:
Variation of Arctic summer sea ice extent over time as depicted in a model simulation. A value of 1.0 on the vertical axis (horizontal line) corresponds to the average Arctic sea ice extent the model simulated for August through October during the twenty year period 1981 to 2000. A year with a value of 0.6 indicates a 40 per cent reduction in sea ice extent compared to the 1981 to 2000 average. Note that by the end of the 21st century, the modeled summer sea ice extent in the Arctic is less than 20 per cent of the 1981 to 2000 average.

1985

RIGHT:
Three model depictions for summertime ice cover (see text for description). White indicates 100 per cent ice cover, deep blue indicates 0 per cent ice cover.

Melting sea ice

Changes in sea ice concentration in the Arctic have been modeled in detail by NOAA's Geophysical Fluid Dynamics Laboratory for the period 1861-2100. The feature shown in the figures represented here is the sea ice concentration simulated over August through October of each year – the months of the year when Northern hemisphere sea ice concentrations generally are at a minimum. Three years (1885, 1985 and 2085) are shown to illustrate the trend simulated by the model. A clear reduction of sea ice extent over time is visible, with the rate of decrease being greatest during the 21st century portion of the simulations.

2085

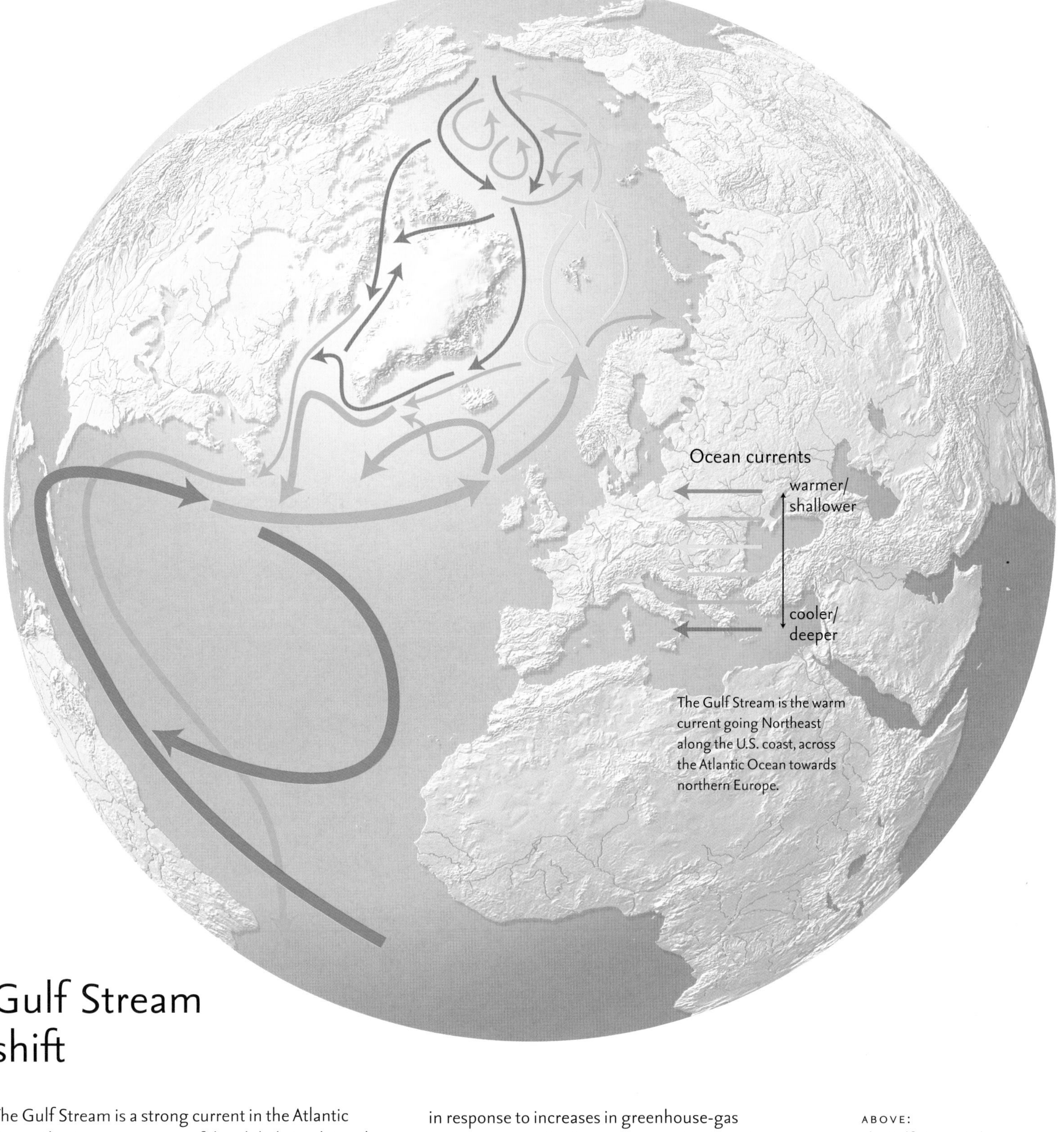

Ocean currents
warmer/
shallower
cooler/
deeper

The Gulf Stream is the warm current going Northeast along the U.S. coast, across the Atlantic Ocean towards northern Europe.

Gulf Stream shift

The Gulf Stream is a strong current in the Atlantic Ocean that is a component of the global circulation by which heat and water are transported throughout the world's oceans, contributing to a process known as the "meridional overturning circulation" (MOC). One of the current hypotheses for future abrupt climate change relates to a slowdown of the Atlantic component of the MOC, since its large net northward heat transport helps to regulate northern hemisphere temperatures. Recent ocean observations have indicated that there is insufficient evidence for a decreasing trend in the rate of Atlantic MOC over the past several decades. However, The IPCC (2007) reports that simulations with coupled ocean-atmosphere climate models suggest a weakening of the Atlantic MOC of up to 60 per cent by the year 2100 in response to increases in greenhouse-gas concentrations. Such a change would affect climate patterns in large portions of our world.

Factors that will affect the magnitude of the MOC and slow the overturning, include the amount of fresh water introduced in the North Atlantic, as a result of melting of Arctic ice, as well as the warming of surface waters from general global climate change.

ABOVE:
The Gulf Stream is the warm surface component shown paralleling the U.S. East Coast. Its warm waters are partially directed to the North Atlantic offshore of the United Kingdom, Greenland and Iceland, where cooling temperatures cause the waters to become more dense, sink, and continue their circulation at depth southward.

Changing sea levels

Throughout geological time, worldwide sea levels have risen and fallen across the millennia. During the past 1,000 years, where the paleoclimate record is most complete, we know that the overall world climate has been relatively constant. In the past 100+ years, we have seen changes in these long-term processes which are incongruent with the geological cycle. As the earth continues to warm, global sea level has risen by 0.5–0.8 in (1.2–2 cm) per decade over the last century.

Any acceleration in the rate of relative sea level rise is cause for concern across the globe, but most especially in low-lying areas where subsidence and erosion problems already exist. Among the range of anticipated impacts are: accelerated erosion; higher/deeper and more frequent flooding of wetlands and other low lying areas; increased frequency of flooding due to more intense storm surges; increased wave energy in nearshore areas; upland and landward migration of wetlands, beaches and other habitats where possible (that is, not inhibited by roadways or other such "barriers"); groundwater intrusion into coastal freshwater aquifers; and damage to a wide range of coastal infrastructures.

Accelerated rates of sea-level rise due to human-induced climate change have caused concern for coastal areas since the issue emerged more than twenty years ago. Less appreciated is the fact that even if the climate is stabilized, sea levels will continue to rise for many centuries due to the long timescales of the oceans and the large ice sheets. Rapid sea-level rise—greater than 3.3 ft (1 m) per century—raises the most concern as it is commonly felt that this may overwhelm the capability of coastal societies to respond and lead to a widespread forced coastal retreat in many areas with accompanying large economic losses. Regionally, most threatened land is in North America, central Asia, and insular nations. In terms of population, east and south Asia will experience significant challenges due to their large populated delta areas.

Within the U.S., islands, the Arctic, and coastal Louisiana are generally regarded as at most risk. The northwestern Hawaiian Islands are of particular environmental concern because of their high conservation value, the high concentration of endemic, endangered and threatened species, and large numbers of nesting seabirds.

The three regions which are generally thought to be the most vulnerable to sea level rise, i.e. the Pacific, Indian Ocean, and Caribbean islands, will only bear a portion of the total global damage due to their relatively small populations and less-developed shorelines. It is felt that many of these areas will simply be largely inundated in the absence of substantial investments in shoreline protection. While various methods of "protection" have been discussed, it is thought that many such options will be cost-prohibitive for all except the large industrialized economies. Actual adaptation responses to sea-level rise are complex, dependent upon economic growth as well as the relative rate of sea-level rise. Historical experience shows that most protection has been a reaction to actual or near disaster; there are few case studies of anticipatory protection outside of the Netherlands. A cycle of decline in some coastal areas may occur, especially in a world where capital is highly mobile and collective action weak. A most interesting area to watch will be that of China, due to the significant ongoing economic growth in its coastal regions.

RIGHT, BELOW AND NEXT PAGE:
The Maldive islands are under particular threat from sea level rise. Most of these islands lie no more than a few meters above sea level. Many of the smaller islands are already submerged and further sea level rise may wipe this country off the map within a hundred years.

RIGHT:
A new coastline for Florida if sea levels rise by 23ft (7 m)? Computer models give us some very specific ideas of what lies in store at different degrees of temperature change. Two degrees is the threshold above which the Greenland ice sheet will melt irreversibly, according to several studies. This would eventually raise global sea levels by 23 ft (7m), flooding coastal cities such as London, Miami, New York, and Bangkok. Most glaciologists think that Greenland will take several centuries to melt completely however.

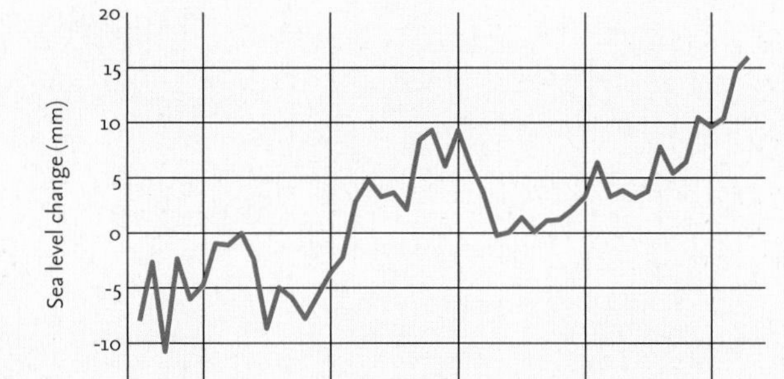

BELOW:
The tiny green islands of the Maldives surrounded by the Indian Ocean.

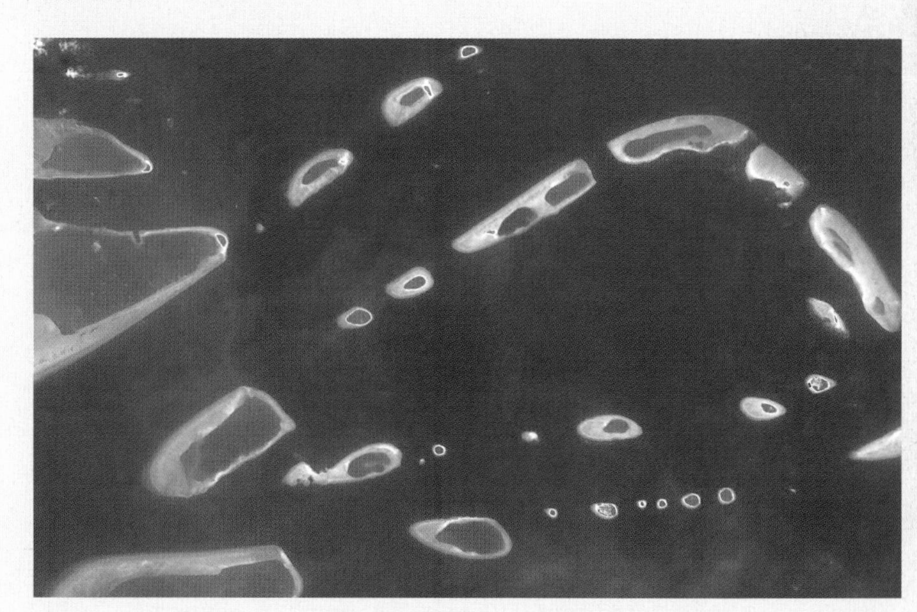

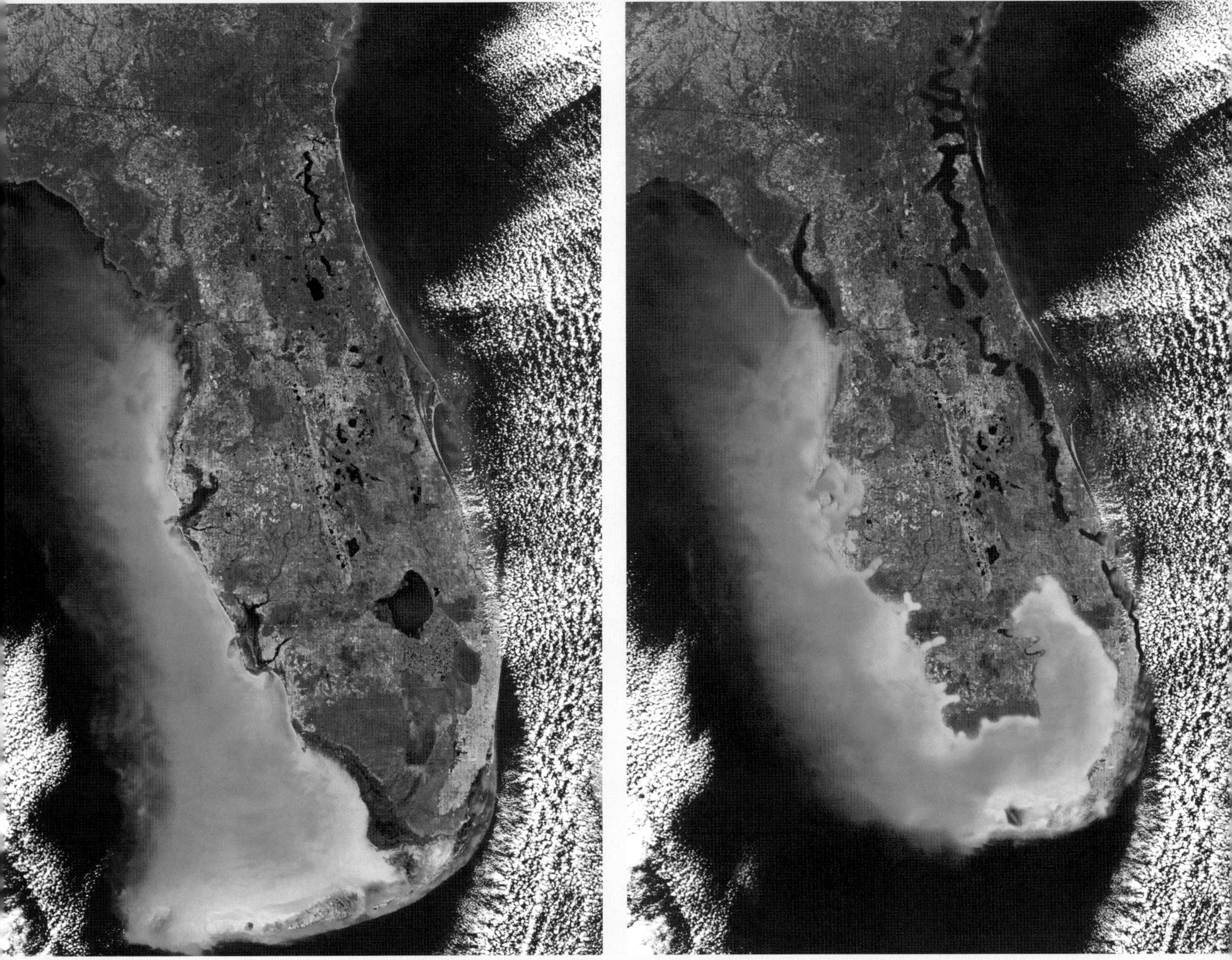

Cities at risk

Since the beginning of human settlements, we have settled at the edge where land and water converge as our food, transportation and waste disposal needs were easily met. More recently in the U.S., the rate of relative population growth along the coast has soared as a result of an expanding coastal recreation and tourism economic sector. In the past fifty years, the density and the economic value of the built environment has escalated. Today, the U.S. coastline, while comprising about 20 per cent of the nation's land holdings, houses over 50 per cent of the U.S. population and generates nearly 60 per cent of the U.S. economy (www.oceaneconomics.org).

Coastal populations around the world are also growing at a phenomenal pace. Already, nearly 66 per cent of the world's population—almost 3.6 billion people—live on or within 100 miles (160km) of the coast. In three decades, it is estimated that 6 billion people will live along coasts—nearly 75 per cent of the world's population. In much of the developing world, coastal populations are exploding. Two out of three cities with populations of 2.5 million or more are located along coasts in southeast Asia, and of the 77 major cities in Latin America, 57 are built in coastal areas.

With the threat of sea level rise, many of the world's coastal cities and towns will have to take measures to ensure their continued survival. For some though, this may be impossible.

ABOVE:
Male, the capital of the Maldives, just 6.5 ft (2m) above sea level. Its reclaimed land is however, lower. After storms in 1987 and 1988 flooded the reclaimed areas, a series of break-waters were built on the outer coast to protect the town. However, they will not prevent flooding from a sustained rise in sea level.

ABOVE RIGHT:
In the event of sea level rise, the world's flood defences will be under increased pressure. This computer generated image depicts a scenario where the Thames flood barrier (see page 113) is breached during a higher than average high tide and storm surge. (From the film Flood)

Threat of rising sea level

- Major cities at risk of submersion
- Coastal areas at greatest risk
- Islands and archipelagos
- Areas of low-lying islands

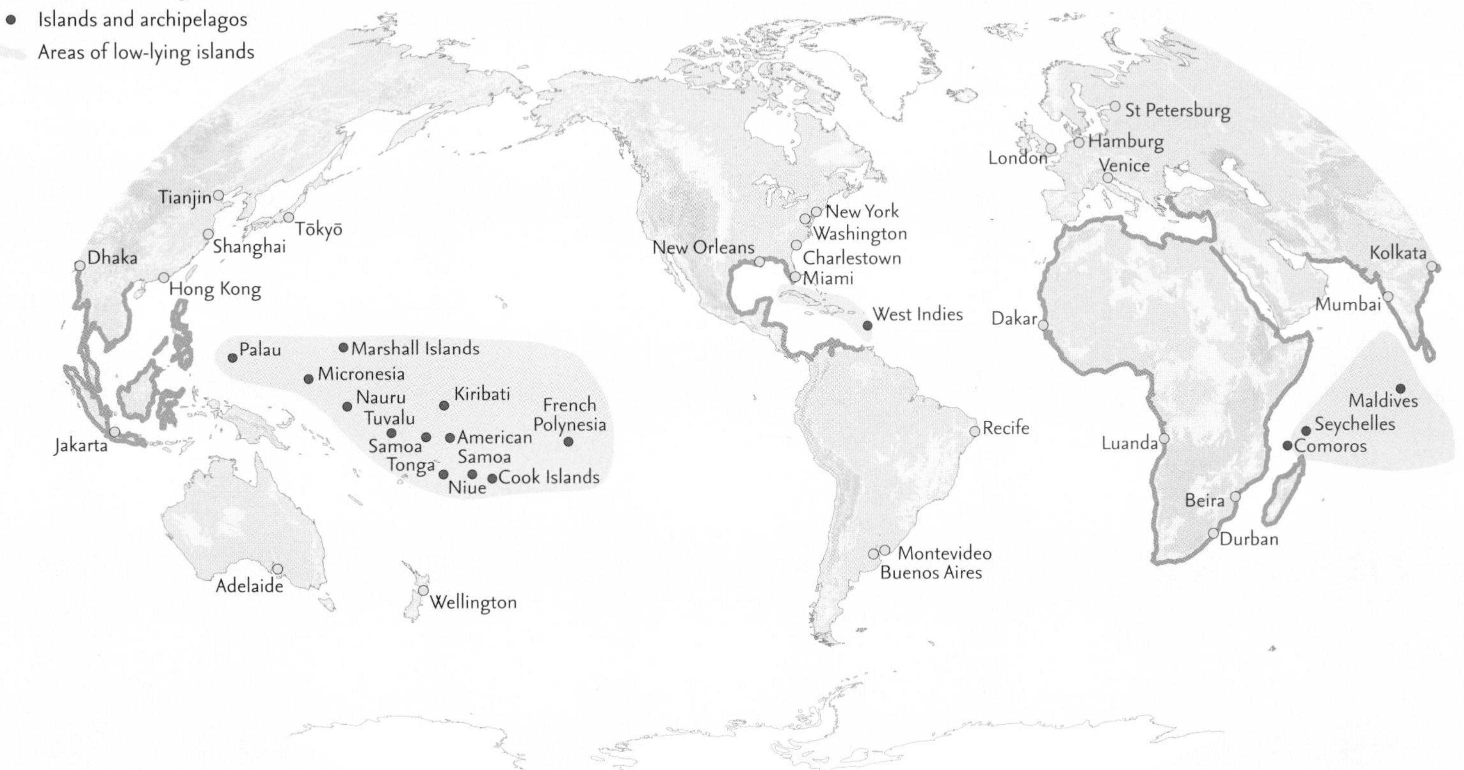

Afterword

Richard Spinrad
Albert E. Theberge

There are many ways of describing the ocean – vast, inspiring, terrifying, brooding.... Before the advent of nineteenth century sounding technology, the ocean was described as "unfathomable," meaning unable to have its depth measured. However, the ocean remains unfathomable in the larger sense that we may never know all there is to know about the ocean, its life, and its processes. Will we ever view in their entirety the grand vistas of earth's great undersea mountain ranges and deepest depressions? Because we are curious and driven to explore, we have found ways to plumb the depths, measure the characteristics of the waters, and observe and describe the creatures of the sea from estuary to abyss. In the early years of ocean science, much of this work was done piecemeal – marine biologists categorized the life of the sea, marine geologists studied sediments and tried to visualize the configuration of the seafloor from sparse soundings, and physical oceanographers attempted to comprehend the driving forces and nature of the small and great movements of the waters from woefully inadequate samples and observations. Because of the disjointed nature of these investigations, it became apparent that to understand the ocean involved cooperation and communication between scientists from many disciplines.

This view was expressed eloquently one hundred years ago (in 1908) by the editors of the first issue of the *Internationale Revue der gesamten Hydrobiologie und Hydrographie* who wrote of "... the necessity of a synthesis of our biological and hydrographical-geological knowledge of the waters. These two spheres of investigation are inseparable, since the water, whether as river, lake, or sea, is never a factor in the shaping of the earth without being also a medium for life, and, on the other hand, is never a medium for life without at the same time having an important influence in the shaping of the earth's surface." Since that time, four generations of ocean scientists have worked to bring about this synthesis in a process that will continue into the foreseeable future. Along the way, the recognition of the profound interconnections between the ocean and the atmosphere has also come to the forefront of the oceanic and atmospheric sciences.

This recognition of the inherent and complicated connections between all sciences of the sea is as valid now as it was a century ago, and will be a mainstay of our continued understanding of the ocean world. It is through this ever-improving awareness of the way in which biology, chemistry and physics interact, as well as our growing recognition of the human influence on these processes, that we will ensure for our oceans a balanced state of economic development, environmental stewardship, and security: protecting lives, livelihoods, and our quality of life!

ABOVE:
One hundred and fifty years ago the great English naturalist Edward Forbes, a pioneer of modern oceanography, included the cartoon above as the frontispiece to "The Natural History of the European Seas". Through this image, he shared his vision of the joy and wonder of ocean exploration and discovery. Past and present generations of ocean scientists have been guided by the example of Edward Forbes and the other pioneers of oceanography. Future generations of ocean scientists will continue sharing his vision as they strive to unlock the secrets of the sea.

Glossary

Aphelion
point of orbit of a planet or satellite where it is furthest from the sun

Apogee
point of orbit of a planet or satellite where it is furthest from the earth

Asthenosphere
lower part of the upper mantle of the earth's surface, made of soft molten rock about 62–217 miles (100–350 km) beneath the surface

Atmospheric circulation
large scale flow of air circulating around the earth

Atoll
a circular or horseshoe array of low-lying coral islands, based on a volcanic seamount, which surrounds a lagoon

Avulsion
sudden erosion or loss of land through water action

Barrier island
long, narrow island of sand and sediment, separated by a lagoon or salt marsh. Susceptible to weather and tides

Barrier reef
long, narrow coral reef parallel to shore separated by a lagoon

Benthic zone
area of the ocean near to the ocean floor

Black smoker
type of hydrothermal vent found on the ocean floor which emits plumes of very hot, mineral rich fluid

Blowhole
vertical vent from a cave roof to the outside which blows water up it due to the pressure of the incoming tide

Chemosynthesis
process whereby chemical energy is used to sustain a food chain as opposed to light energy (photosynthesis)

Continental drift
previously held theory of land masses moving over millions of years, having once been joined in one mass called Pangaea

Cyclone
area of low pressure around which air circulates clockwise in southern hemisphere and counterclockwise in northern hemisphere

Delta
fan shaped area at a river mouth where eroded material is deposited and not washed away by ocean currents

Deposition
process of material dropped by water

Doldrums
area of calm or light winds near the equator. Weather is sultry and sailing ships can be becalmed for days

Dune
hill or ridge of sand created by wind

Earthquake
shock waves which travel through the earth and across the surface due to sudden release of energy in the earth's crust

Ecosystem
ecological community including non-living elements

El Niño
occasional environmental phenomenon, characterized by, among other things, unusual warming of the tropical areas of the Pacific Ocean

Erosion
wearing away of land naturally by wind, water, or ice. For example, the Grand Canyon and Yosemite are products of erosion by water and ice respectively

Extra-tropical cyclone
cyclones whose diameter can stretch for thousands of miles

Geologic time
relative time scale based on fossil content

Global warming
certain natural and human-produced gases prevent the sun's energy from escaping back to space leading to an overall rise in the temperature of the Earth's atmosphere

Gulf Stream
warm ocean current in north Atlantic stretching from mid-Atlantic United States to northern Europe

Gyre
a large system of circulating water currents

Horse latitudes
subtropical high pressure areas, 30–35° north and south where wind is light and variable. Weather is hot and dry

Hurricane
intense tropical cyclone in north Atlantic and east and central Pacific where winds spiral inwards to a core of low pressure

Hydrologic cycle
the movement of water from earth, to atmosphere and back to earth, in gas, liquid or solid form

Hydrothermal vent
cold seawater heated by underlying magma in a crack of the seafloor, bubbles up as hot or warm springs

Hypoxic zone
area of the ocean low in oxygen so cannot support or stresses, marine life

La Niña
occasional environmental phenomenon, characterized by, among other things, unusual cooling of the tropical areas of the Pacific Ocean

Lithosphere
outer layer of the earth. Includes crust and upper mantle measuring 50–62 miles (80–100 km) thick

Longshore drift
movement of sand and other material along a beach due to waves breaking on the shore at an angle

Magma
molten rock formed within the earth. Can be liquid or a mixture of liquid, crystals, and dissolved gases

Marine terrace
forms where a sea cliff, with a wave cut platform before it, is raised above sea level

Monsoon
seasonal weather patterns caused by the effects of differential heating first recognized in India and southern Asia

Ocean conveyor belt
global current which links the Pacific, Atlantic, Indian, and Southern Oceans. Also called Thermohaline Circulation

Oceanic circulation
movement of water within the oceans mainly driven by wind, or differences in temperature or salinity

Oceanography
study of the oceans and their ecosystems, and chemical and physical processes

Pelagic zone
open water areas of the ocean

Perigee
point of orbit nearest to earth of the moon or a satellite

Perihelion
point of orbit nearest to the sun of a planet or a satellite

Plankton
small floating or slow moving aquatic plant (phytoplankton) or animal (zooplankton)

Plate tectonics
the concept that the earth's crust consists of large, rigid but floating plates which cause geologic activity where they meet

Precipitation
any form of moisture deposited from the atmosphere

Raised beach
wave cut platform which has been raised above sea level by geological movement or left after a fall in sea level

Reef
ridge of rocks, coral or sand rising from the seabed which reaches to just below the water level. May be exposed at low tide

Salinity
the concentration by weight, of salts in water

Salt marsh
marsh area flooded by sea water

Sandbar
submerged or partly exposed ridge of sand and sediment created by wave action off a coast

Sea breeze
onshore wind caused by temperature differences between land and sea

Sea fog
air from a warm water area meets colder water, causes condensation, forming low cloud near the ocean surface

Sea level
point from which land heights are taken. Mean sea level is the average level taken from high and low tides

Sea smoke
also called steam fog – when very cold polar air meets warmer ocean areas condensation produces this type of fog

Sea surface temperature
temperature of sea to about one foot's depth

Seamount
mountain on the seabed, usually conical and of volcanic origin

Stack
isolated pillar of rock separated from cliff by erosion

Storm surge
abnormally high water level on the coast due to wind on the water surface

Subduction
process of one tectonic plate forced down and under another, after collision

Surface current
horizontal movement of water of the upper layer of the ocean waters e.g. Gulf Stream

Swell
the extent of wind-formed waves after the wind has died down

Trade winds
easterly wind (from the east) heading towards the equator

Typhoon
tropical cyclone exceeding 64 knots (119 km/hr) which occurs in western Pacific

Volcano
point where magma is forced up a weak fissure in the earth's crust, releasing molten lava, ash, and gases

Wave cut platform
horizontal rock surface at a cliff base formed by wave action

Westerlies
winds from the west between 30° and 50° latitude, northern and southern hemispheres

Organizations

ASEAN
Association of South East Asian Nations

GEO
Group on Earth Observations

IUCN
World Conservation Union (formerly known as International Union for Conservation of Nature and Natural Resources)

INA
Institute of Nautical Archaeology

IPCC
Intergovernmental Panel on Climate Change

IWC
International Whaling Commission

JTWC
Joint Typhoon Warning Center

MMS
Minerals Management Service

NATO
North Atlantic Treaty Organization

NMSP
National Marine Sanctuary Program

NOAA
National Oceanic and Atmospheric Administration

RSMC
Regional Specialized Meteorological Center

TCWC
Tropical Cyclone Warning Center

UNCLOS
United Nations Convention on the Law of the Sea

UNESCO
United Nations Educational, Scientific, and Cultural Organization

WCPA
World Commission on Protected Areas

Index

Page numbers in **bold** refer to captions and illustrations.

Acknowledgments

Design, origination, and editorial control by Collins Geo, HarperCollins Publishers, Glasgow, UK

Authors and Contributors

Coordination of authors and NOAA editorial input: Captain Albert E. Theberge, NOAA Corps (ret.), NOAA Central Library

Foreword
Author: Vice Admiral Conrad C. Lautenbacher, Jr., U.S. Navy (Retired) Undersecretary of Commerce for Oceans and Atmosphere and NOAA Administrator

The ocean floor
Author: Dr. Robert W. Embley, NOAA Research

Ocean zones
Authors: Dr. Dwayne Meadows and Dr. Marie-Christine Aquarone, NOAA National Marine Fisheries Service

Water and seawater
Author: Lt. (j.g.) Patrick L. Murphy, NOAA Corps, NOAA National Marine Fisheries Service

Water movement and circulation
Lead author: Dr. Rick Lumpkin, NOAA Research, Atlantic Oceanographic and Meteorological Laboratory
Tides and *Mean sea level*: Dr. Stephen Gill, Chief Scientist, NOAA Center for Operational Oceanographic Products and Services
Waves: Dennis Cain, NOAA National Weather Service

Oceans and the climate
Lead author: Michael Johnson, NOAA Climate Program Office
El Niño and La Niña: Dr. Michael J. McPhaden, NOAA Research, Pacific Marine Environmental Laboratory

Oceans and the weather
Lead author: Hugh Cobb, NOAA National Weather Service, National Hurricane Center
Extra-tropical cyclones, *Sea fog*, *Sea smoke*, and *Sea breezes*: Brett Lutz, NOAA National Weather Service
Technical review: John L. Beven, Hurricane Specialist, NOAA National Weather Service, National Hurricane Center

Dynamic coasts
Lead author: Lt. (j.g.) Patrick L. Murphy, NOAA Corps, NOAA National Marine Fisheries Service
Beaches and barrier islands: Lt. (j.g.) Mark Blankenship, NOAA Corps, National Ocean Service

Natural hazards
Lead authors: Helen Wood, Senior Advisor for NOAA Satellites and Information and Margaret Davidson, Director NOAA Coastal Services Center
Tsunamis: Dr. Eddie Bernard, NOAA Research, Director Pacific Marine Environmental Laboratory
Storm surge, *Rogue waves*, and *Rip currents*: Dennis Cain, NOAA National Weather Service
Waterspouts: NOAA National Weather Service Miami – South Florida Forecast Office
Biological hazards in the sea: Captain Albert E. Theberge, NOAA Corps (ret.), NOAA Satellites and Information

The poles
Lead author: Dr. Pablo Clemente-Colón, NOAA, Satellite and Information Service, Chief Scientist National Ice Center
Arctic Ocean human impacts: Dr. Kathleen Crane, NOAA Research, Arctic Research Program
Antarctica human impacts: Dr. Christopher D. Jones, NOAA National Marine Fisheries Service

Vital ecosystems
Lead author: Dr. Teresa McTigue, NOAA Center for Coastal Monitoring and Assessment
Large marine ecosystems: Dr. Kenneth Sherman, NOAA National Marine Fisheries Service, Narragansett Laboratory
Salt marshes: Alisha Goldberg, College of the Mainland
Mangroves: Felicity Burrows, The Nature Conservancy Bahamas Program
Estuaries: Dr. Susan White, NOAA Estuarine Reserves Division
Sea grass habitat: Dr. Gordon W. Thayer, NOAA National Ocean Service (ret.)
Kelp forests: Dr. Katrin Iken, School of Fisheries and Ocean Sciences, University of Alaska, Fairbanks
Coral reefs: Liza Johnson and Alissa Barron, NOAA Coral Reef Conservation Program

Exploring the oceans
Lead author: Captain Albert E. Theberge, NOAA Corps (ret.), NOAA Satellites and Information
Exploration technology: Gretchen Imahori, NOAA National Ocean Service

Human interaction
Lead authors: Craig N. McLean, Deputy Assistant Administrator, NOAA Research and Anne M. Readel NOAA Research
Ocean observation: Zdenka Willis, NOAA Director Integrated Ocean Observing System (IOOS) Program Office
Marine biology and Nature and industry sharing common ground: Emma Hickerson, NOAA National Ocean Service, Sanctuary Research Coordinator
Telecommunication cables and Energy: Anne M. Readel, NOAA Research
Sea ports, *Shipping routes*, and *Commerce*: Captain Steven Barnum, NOAA Corps, Hydrographer of the United States, NOAA National Ocean Service, Director Office of Coast Survey; Richard Edwing, Meredith Westington, Ashley Chappell, and John Nyberg, NOAA National Ocean Service
Predicting shipwreck locations: Bruce Terrell, Chief Marine Archaeologist, NOAA National Ocean Service
Fishing and fishing grounds: Laura Oremland, NOAA National Marine Fisheries Service
The Baltic Sea: Joseph R. Vadus, NOAA National Ocean Service (ret.)
Sea floor minerals: Dr. John C. Wiltshire, NOAA Undersea Research Center for Hawaii and the Western Pacific
International law of the sea: Meredith Westington, NOAA National Ocean Service
The South China Sea: Meredith Westington, NOAA National Ocean Service
Diving and Submersible technology: John J. McDonough, NOAA Research, Office of Ocean Exploration
Pollution, *Marine debris*, and *Oil*: Dr. Amy Merten, NOAA National Ocean Service; Kimberly Newman, Coastal Response Research Center, University of New Hampshire; Captain Ken Barton, NOAA Corps, National Ocean Service, and Doug Helton, NOAA National Ocean Service
Ancient shipwrecks of the Mediterranean and Black Seas: Dr. Dwight Coleman, Assistant Director University of Rhode Island Inner Space Research Center
Cultural history and traditions: Carol Bernthal, NOAA National Ocean Service

Conservation
Lead author: Joseph Uravitch, Director, NOAA National Marine Protected Areas Center
Worldwide marine protected areas: Annie Hillary, NOAA National Ocean Service
Coral reef conservation: Alissa Barron, NOAA Coral Reef Conservation Program, Tyler Christensen and Alan J. Strong, NOAA Satellite Oceanography Division
Sustainable fisheries: Gordon J. Helm, NOAA National Marine Fisheries Service
Protection of marine mammals and turtles: Trevor R. Spradlin and Kristy Jayne Long, NOAA National Marine Fisheries Service
The National Marine Sanctuary System: Michael T. Murphy, NOAA National Ocean Service

The future
Lead authors: Dr. Richard Spinrad, Assistant Administrator NOAA Research, and Margaret Davidson, Director NOAA Coastal Services Center
Future of fish: Dr. Richard Methot, NOAA National Marine Fisheries Service

Afterword
Dr. Richard Spinrad, Assistant Administrator NOAA Research, and Captain Albert E. Theberge, NOAA Corps (ret.), NOAA Satellites and Information

Images

Maps and diagrams compiled by Collins Geo with input from NOAA: pages 12–13, 14, 17, 18–19, 23, 44, 56–57, 71, 78, 79, 97, 108, 109, 131, 132–133, 139, 153, 179, 187, 188, 203, 207, 227, 241, 242, 243, 247.

All images courtesy of NOAA unless stated otherwise.

Key: c = center; t = top; b = bottom; l = left; r = right

8-9 Geo-Innovations/Alan Collinson Design; **10 and 11** Image produced by Dr. Walter H.F. Smith, NOAA Satellites and Information, Center for Satellite Applications and Research; **15** Walter Smith and David Sandwell, NOAA; **16** t and b: Walter Smith and David Sandwell, NOAA; **20-21** © Jonathan Blair/Corbis; **23** t: José F. Vigil, USGS; **27** c: Image courtesy of IFE, URI-IAO, UW, Lost City science party, NOAA; **30-31** NASA/Science Photo Library; **33** rtc: W Haxby, Lamont-Doherty Earth Observatory/Science Photo Library; **35** t and b: Sarah Alice Lee; **37** b: T. W. Pietsch/University of Washington; **38 and 39** Images courtesy NASA/GSFC/LaRC/JPL, MISR Team; **41** NASA/MODIS; **42** Claire Ting/Science Photo Library; **45** b: Image courtesy of NASA/GSFC/MITI/ERSDAC/JAROS, and US/Japan ASTER Science Team; **47** Douglas Faulkner/Science Photo Library; **48-49** MODIS/NASA; **50-51** NASA/Science Photo Library; **53** t and b: NASA/MODIS; **54-55** Data source: National Centers for Environmental Prediction (NCEP) Climate Diagnostics Center, Boulder, CO; **56-57** Data source: Global Drifter Program; **58-59** NASA and Science Photo Library; **60 and 61** Andrew J Martinez/Science Photo Library; **63** t: Fleur Gayet; **63** b: Mark Steward; **64-65** NASA/MODIS; **67** t: Dr Keith Wheeler/Alamy; **67** b: © Visual&Written SL/Alamy; **68** t: SeaWiFS/GeoEye; **68** b: Copyright © 2007 EUMETSAT; **70-71** NOAA/GOES; **72** TOPEX/Poseidon, NASA JPL; **73** t: © Bill Brooks/Alamy; **73** cr and br: Image courtesy of USGS; **74-75** Mark Steward; **77** NASA; **80-81** Sergio Pitamitz/Alamy; **83** bl: Image courtesy of Jeff Halverson, TRMM Outreach Scientist and Hal Pierce, TRMM Visualizer, NASA GSFC; **86** NASA; **87** SC Photos/Alamy; **88** l: Paolo Koch/Science Photo Library; **88** r: Brian Brake/Science Photo Library; **89** J Marshall-Tribaleye Images/Alamy; **90** Peter Menzel/Science Photo Library; **91** t: © Chris Linder/Alamy; **91** b: © David Wall/Alamy; **92-93** Tony Craddock/Science Photo Library; **95** t: © CORBIS; **95** b: Arctic Images/Alamy; **96** t and b: Image courtesy of USGS; **98-99** Dirk Wiersma/Science Photo Library; **101** tl and tr: Image courtesy of USGS; **102** and **103** Images reproduced by kind permission of UNEP; **104-105** Alexis Rosenfeld/Science Photo Library; **108** b: Photographer unknown; **109** tl and tr: IKONOS images © CRISP 2004; **110 and 111** IKONOS images © CRISP 2004; **113** t: Skyscan/Science Photo Library; **114** t: Alan Sirulnikoff/Science Photo Library; **115** t and b: Michel Gunther/Still Pictures; **118 and 119** MODIS/NASA; **124** t: National Snow and Ice Data Center, Boulder, CO, U.S.A.; **125** MODIS/NASA; **127** br: © Greenpeace/Cunningham; **128** br: National Snow and Ice Data Center, Boulder, CO, U.S.A.; **134-135** Bernhard Edmaier/Science Photo Library; **137** Bernhard Edmaier/Science Photo Library; **138** Image courtesy NASA/GSFC/LaRC/JPL, MISR Team; **140** Kaj R Svensson/Science Photo Library; **141** Earth Satellite Corporation/Science Photo Library; **143** NASA/GSFC; **145** t: Adam Jones/Science Photo Library; **146** Alexis Rosenfeld/Science Photo Library; **148** Alexis Rosenfeld/Science Photo Library; **150** Bud Lehnhausen/Science Photo Library; **151** Gregory Ochocki/Science Photo Library; **162** tc and tr: Oceanographic Museum of Monaco (NOAA); **162** b: North Wind Picture Archive/Alamy; **172** t: Oceanographic Museum of Monaco (NOAA); **174-175** Sarah Alice Lee; **178** b: European Space Agency/Science Photo Library; **182** t: FGBNMS/E L Hickerson; **182** b: FGBNMS/D C Weaver; **183** t: FGBNMS/E L Hickerson; **185** t: Phillip Hayson/Science Photo Library; **190** l: United States Navy Historical Center and University of New Hampshire Center for Coastal and Ocean Mapping/ Joint Hydrographic Center; **191** Peter Scoones/Science Photo Library; **192** l and r: Institute for Exploration and Institute for Archaeological Oceanography, University of Rhode Island; **193** t: © Jonathan Blair/CORBIS; **193** b: Institute for Exploration and Institute for Archaeological Oceanography, University of Rhode Island; **194** Mark Steward; **195** t: Mark Steward; **195** b: Mark Zylber/Alamy; **198** Peter Bowater/Alamy; **199** t: Image supplied courtesy of Global Marine Systems Limited; **199** b: Gary Hincks/Science Photo Library; **200** b: Charles D Winters/Science Photo Library; **203** t: TeleGeography Research; **203** b: Image supplied courtesy of Global Marine Systems Limited; **204** US Coast Guard; **205** t: NASA/MODIS; **205** bl: Classic Image/Alamy; **206** IKONOS image courtesy of GeoEye; **207** t: Peter Bowater/Alamy; **208** NASA/MODIS; **209** NASA/MODIS; **210** t: Peter Adams Photography/Alamy; **212-213** Georgette Douwma/Science Photo Library; **215** t: Fred McConnaughey/Science Photo Library; **215** br: Pat & Tom Leeson/Science Photo Library; **216** r: National Estuarine Research Reserve System, NOAA; **218-219** t: Alexis Rosenfeld/Science Photo Library; **219** b: Peter Scoones/Science Photo Library; **222** t: British Antarctic Survey/Science Photo Library; **223** b: Peter Menzel/Science Photo Library; **224-225** t: Peter Scoones/Science Photo Library; **228-229** Christian Darkin/Science Photo Library; **232** IKONOS images courtesy of GeoEye; **233** Simon Fraser/Science Photo Library; **234** Andrew J Martinez/Science Photo Library; **235** t: Volker Steger/Science Photo Library; **236-237** Alexis Rosenfeld/Science Photo Library; **238** t: Martin Bond/Science Photo Library; **239** Dr Keith Wheeler/Science Photo Library; **240** NASA; **241** Globe: Geo-Innovations/Alan Collinson Design; **243** Globe: Geo-Innovations/Alan Collinson Design; **244** NASA/ASTER; **245** t: Jack Sullivan/Alamy; **245** bl and br: NASA; **246** Shahee Ilyas; **247** Courtesy of Power

BOCA RATON PUBLIC LIBRARY, FLORIDA
3 3656 0450650 0

Collins